IMAGES
of America

TALENT

ON THE COVER: A group of talented fellows who played musical instruments formed the Talent City Cornet Band, pictured here in 1909. They highlighted public and private events with memorable music. Because their sound was not recorded, it lasted only in the minds of the players and the crowds that gathered to hear them. (Talent Historical Society.)

IMAGES
of America

TALENT

Jan Wright of the
Talent Historical Society

ARCADIA
PUBLISHING

ISBN 978-0-7385-7146-1

Published by Arcadia Publishing
Charleston SC, Chicago IL, Portsmouth NH, San Francisco CA

Printed in the United States of America

Library of Congress Control Number: 2008942878

For all general information contact Arcadia Publishing at:
Telephone 843-853-2070
Fax 843-853-0044
E-mail sales@arcadiapublishing.com
For customer service and orders:
Toll-Free 1-888-313-2665

Visit us on the Internet at www.arcadiapublishing.com

To my sister, Char, my north star

CONTENTS

ACKNOWLEDGMENTS

The images in this book did not just show up one day all neatly packaged and captioned. They had been kept by various families and collectors, in some cases for over 100 years, before being brought to light, scanned, and organized in the book now in your hands. If it were possible, each person who left a fingerprint on the photographs would be recognized individually for their contributions along with those who took the time to write names, dates, and places on the back of the photographs. I can only express my gratitude to the whole group. Thanks also go to those who saved diaries, newspapers, letters, scrapbooks, and documents that helped interpret Talent's history and images. The list of contributors includes all the board members of the Talent Historical Society past and present—especially the board president, Robert Casebeer, the City of Talent, and the following individuals who let us rediscover the past through their collections: Richard Allen, Liz Carter, Mike Cotta, Harry "Bud" Gleim, the late Genevieve Holdridge, Robert Jacobs, Ken Kantor, the Kerby family, Jerry Kime, Maxine Kincaid, Richard Lohr, Jim Maize, Jerald McGrew, Ron Ridgway, Terry Skibby, Southern Oregon Historical Society (SOHS), Sandra Stiefel, Everett Taylor, Ben Truwe, James Uerlings, and Ralph Wyant. This has truly been a collaborative effort. Marla Cates, Kay Atwood, Jeff LaLande, and George Kramer paved the way with their historic inventories, interviews, writings, and attention to the built environment for the City of Talent. This project would not have been possible without Sarah Higginbotham, David Gray, and Brian Schick. Kathy Enright and Suzanne M. M. Warner were helpful with the SOHS collection. My sister, Charlotte Wright, and my four children encouraged me through the writing process. My grandchildren inspired me to tell the story of the town of their first memories. Though he has been dead for over 100 years, Welborn Beeson must be credited for his habit of daily record keeping, which contributed to this work and many others. Unless otherwise acknowledged, the photographs in this book are from the Talent Historical Society collection.

INTRODUCTION

The story of Talent, Oregon, is heartbreaking for the Native Americans, who had at least two villages along Wagner and Anderson Creeks, and backbreaking for the white settlers, who liked their land cleared and farmable. In the winter of 1851–1852, a gold rush in southern Oregon attracted the first miners to the area. Their search disrupted the Native American way of life and brought on a war, which pushed most of the Native Americans out of the valley. Some of the miners decided to stay and make a permanent home in addition to the emigrants who came to take up donation land claims in southern Oregon.

The community now called Talent is a portion of the area first referred to as Wagner Creek and formed around a donation land claim settled in 1852 by Jacob Wagner. He was one of the first to settle, and soon after, the creek, the mountain looming to the southwest, the village, and eventually the main business street in town were all named after him. For more than 20 years, the pioneers of the Wagner Creek vicinity had to travel to Phoenix, Ashland, or Jacksonville to pick up their mail, dine at a restaurant, buy dry goods, or get a shave and haircut. Things changed when a man named Aaron Patton Talent came to the area in the 1870s; he bought over 100 acres of the old Wagner land and divided it into lots. The act of selling off small parcels of land allowed a concentration of homes and businesses to form, creating the downtown core of what is now called Talent, Oregon. The town was officially named when the post office opened in 1883 and the population totaled 67 human beings. The community remained small, but the pace of life changed when it offered employment, markets, social opportunities, and places to worship.

One can only imagine what southern Oregon looked like before the settlers began to change the face of the land. There are no known photographic images of the Native American villages or of the first settlements along Wagner Creek. The earliest landscape photographs of the area nearly all include rooftops and fences, but if one looks closely, one can imagine the watersheds; the mountains; the pine, fir, and oak trees; the grasses; and the wildlife that existed for millennia. Bear, Wagner, and Anderson Creeks' flows have been altered by the Talent Irrigation District ditches and by hydraulic mining practices. Deer trails have expanded to footpaths and wagon trails to paved highways. Bridges now cross the creeks. Native plants are harder to find. Cemeteries break up the grassy hills and fences artificially divide the land into angular patterns that replace nature's curves.

The earliest image of the area is a hand-drawn 1854 survey map of Township 38 South Range 1 West, showing the home sites and gardens and, more importantly, the roadways through the area. The Applegate Trail, a southern alternative of the Oregon Trail, followed along Bear Creek through the lands of the Rockfellow brothers, Jacob Wagner, William Smith, and Thomas Chase and followed roughly the same path as the Yreka-Jacksonville road on the map. Pack trains, cattle, and emigrants passed through the Rogue Valley on their way to the Willamette Valley or other settlements to the north. The road to Jacksonville and to Yreka through the Siskiyou Mountains connected the early settlers to the outside world.

Built in 1853 around Jacob Wagner's log cabin, Fort Wagner dominated the valley floor. It was built to protect arriving emigrants and settlers during the Indian Wars of 1853–1856. No battles occurred in Talent, but on several occasions, settlers took shelter inside the fort and some of the wounded soldiers were carried there to recover or die. Although neither drawings nor photographs of Fort Wagner have been found, Martha Hill Gillette described it in *Overland to Oregon* as covering an acre of ground and having 12-foot walls with observation stations at the corners. When the fort fell into disuse, it was torn down. Wagner's cabin remained until 1874, and by 1884, all traces of it, except the mound where the chimney stood, were gone. A drawing of a similar and nearby structure is included in this volume to show the reader what Fort Wagner could have looked like.

Nearly all of the extant photographs of Talent and its citizens document the town's growth after the railroad came through in the 1880s. The advent of the railroad sparked a new growth as strong as the initial wave of emigrants. A business district grew around the tracks, and for the first time residents could shop at home. The much-anticipated depot building arrived only because the Southern Pacific Depot in Medford, considered too small for that town, was moved from Medford by rail in 1900 on a flatbed car and was set in place on Talent turf.

Many of the photographs in this book were taken after 1910, when the city was incorporated and its citizens were eager to brag about it. A favorite spot from which to take photographs was atop the brick schoolhouse on West Wagner Street (where Talent Elementary is now). In 1911, fire destroyed a row of businesses in town, but new buildings were quickly rebuilt to replace them as Talent continued to thrive. By 1912, with a population of 500, a planing mill and box factory; the Bagley Cannery; a Baptist, Christian, and Methodist church; a bank; a new water reservoir; streetlights; a new brick school building; and the first woman in Oregon to be elected to a city position were all in place, with more ambitious undertakings planned for the coming years.

Socially, Talent stood apart from its neighboring communities because of the number of freethinking, progressive citizens who called it home. Distinguished individuals included John Beeson, known for his civil rights advocacy for the Native Americans; Willis J. Dean, known for his superior teaching abilities and intellectual prowess; Welborn Beeson (John Beeson's son), known as the town diarist, surveyor, justice of the peace, and witness to nearly every important activity in the neighborhood; Edward "Eddie" Robison, known as a radical newspaper editor in his youth; and Mr. and Mrs. John Holton, known for their séances. Wagner Creek had its own Universal Mental Liberty Hall, an institution of liberal thinking that invited debate and frequently sparked controversy. The Church of the Brethren, known as Dunkards, had a church house and cemetery in town for its followers. Talent became a populist seedbed in the 1890s but was most well-known for its quirky sentiments toward spiritualism. William Breese had the distinction of being Oregon's first Socialist mayor and was an ardent spiritualist.

From the earliest days of settlement, education was considered important. The first school in the valley, Eden School, was built along Bear Creek off what is now Suncrest Road in Talent. A subscription school, it was organized in 1854 before Jacksonville started the first public school in the county. Scholars also attended school in Gasburg (now Phoenix) at the Colver House until a bigger school for Wagner Creek students was built on the Beeson property. Schools served as social centers, lecture halls, meeting houses, and polling places. The 1899 schoolhouse on Main Street, now the community center, was only in operation as a school for a few years. Population increases made it clear that a bigger building was needed to accommodate the children. In 1911, a brick school, called simply Talent School, was opened and served as a grade school and a high school. A new elementary school replaced the old brick school when it was torn down after much controversy in 1973. The Talent High School continued in what is now the Talent Middle School building until the early 1960s, when it consolidated with Phoenix. On Wagner Creek, a separate school district formed on the Lynch family property and eventually became the Wagner Creek school. Anderson Creek had its own schoolhouse as well.

Agricultural pursuits prevailed as the mainstay of the community for many years. From the valley's first wheat harvest to the first moneymaking orchards, Talent's agricultural reputation was unsurpassed. The 1920s, however, brought a population decrease as a result of the decline in the orchard industry and railroads, the relocation of the Bagley Cannery to Ashland, and the purchase of the State Bank of Talent by the First National Bank of Ashland. Another blow occurred in 1938 when the Pacific Highway through town (now Talent Avenue) was bypassed with a new road (Highway 99), which atrophied the main business district. The Jackson County Poor Farm north of town was removed and replaced by the Education Service District and County Animal Shelter buildings.

Harvesting timber has been a tradition since 1854, when Granville Naylor set up the first sawmill along Wagner Creek. Each time a different owner took over, the sawmill would be moved a little farther up the creek. After World War II, logging operations began to take off. Skeeters Logging Company employed returning soldiers to work in the woods and manage the heavy logging equipment. A timber mill was located along the railroad tracks close to Colver Road where Candlewood Mobile Home Park is now. Logging offered steady jobs for Talent residents through the 1960s, tapering off in the 1970s.

Recreation is a given in the Rogue Valley. Because of the natural beauty in southern Oregon, hiking, camping, fishing, and hunting have been consistent attractions. Many of the images in this volume reflect Talent residents' love affair with the outdoors, automobiles, and each other. Like other Americans, they participated in the ritual of the Sunday drive. Day trips allowed them to explore other parts of the valley and better appreciate their own hometown. Groups, clubs, schools, and lodges all offered social outlets for the community to see, know, and enjoy each other face to face.

In 2010, the Talent Centennial Committee commemorates the 100th year of the city's incorporation with a series of planned events. This volume is one of Talent Historical Society's contributions to that effort. It is hoped that through this book readers will find a new way of looking at their community, enabling them to revive and maintain the community spirit that built this town. Talent Historical Society welcomes further inquiry and contributions to the existing records and photographs of Talent and the upper Bear Creek Valley.

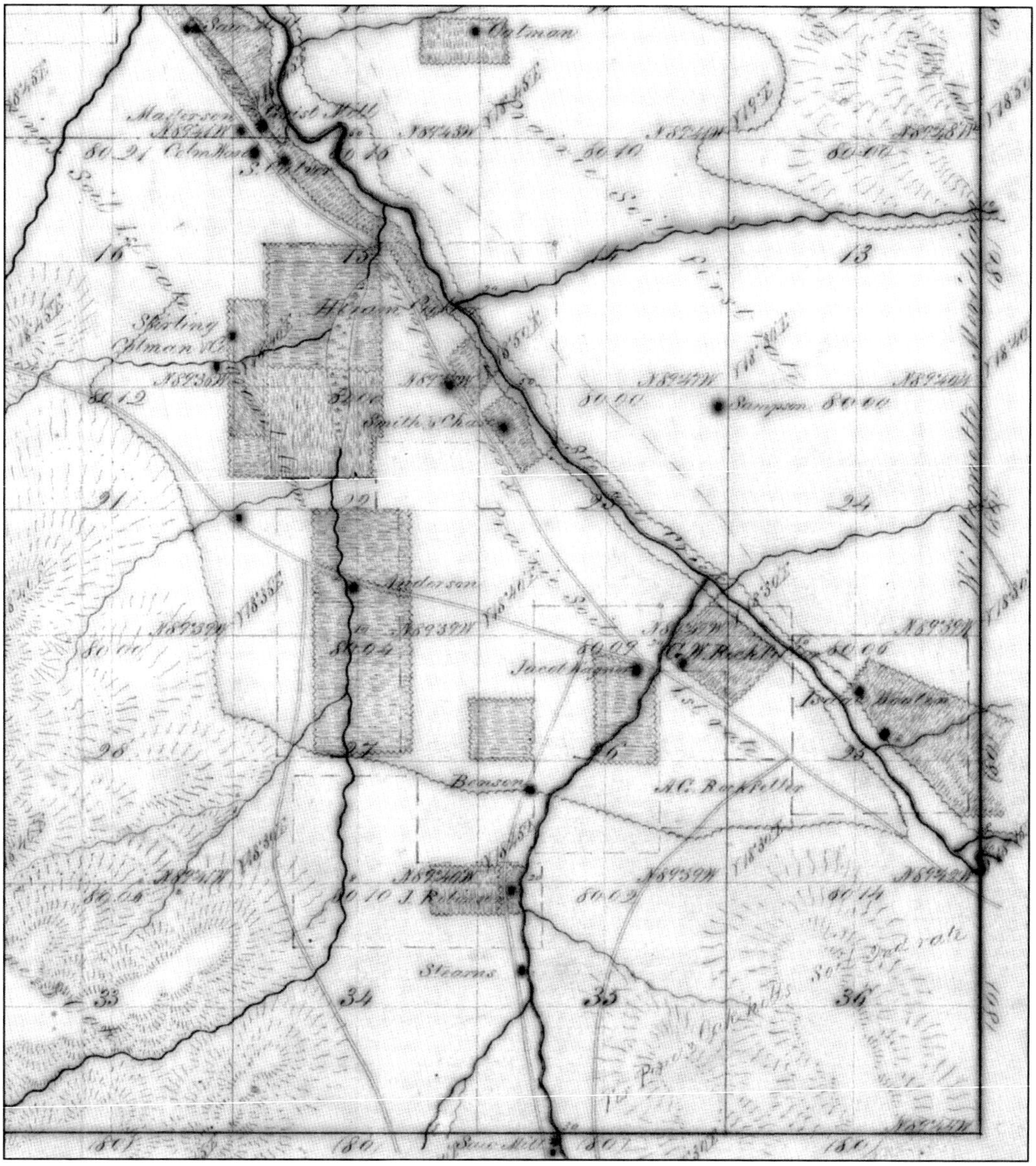

To satisfy government demands, Butler Ives and George Hyde were hired in 1854 to survey Oregon donation land claims. The survey map is the first visual record of township 38 South Range 1 West of the Willamette Meridian. With this map, the names of the first settlers, roads, and creeks are forever imbedded on the pages of history. (Bureau of Land Management.)

One

LANDSCAPE

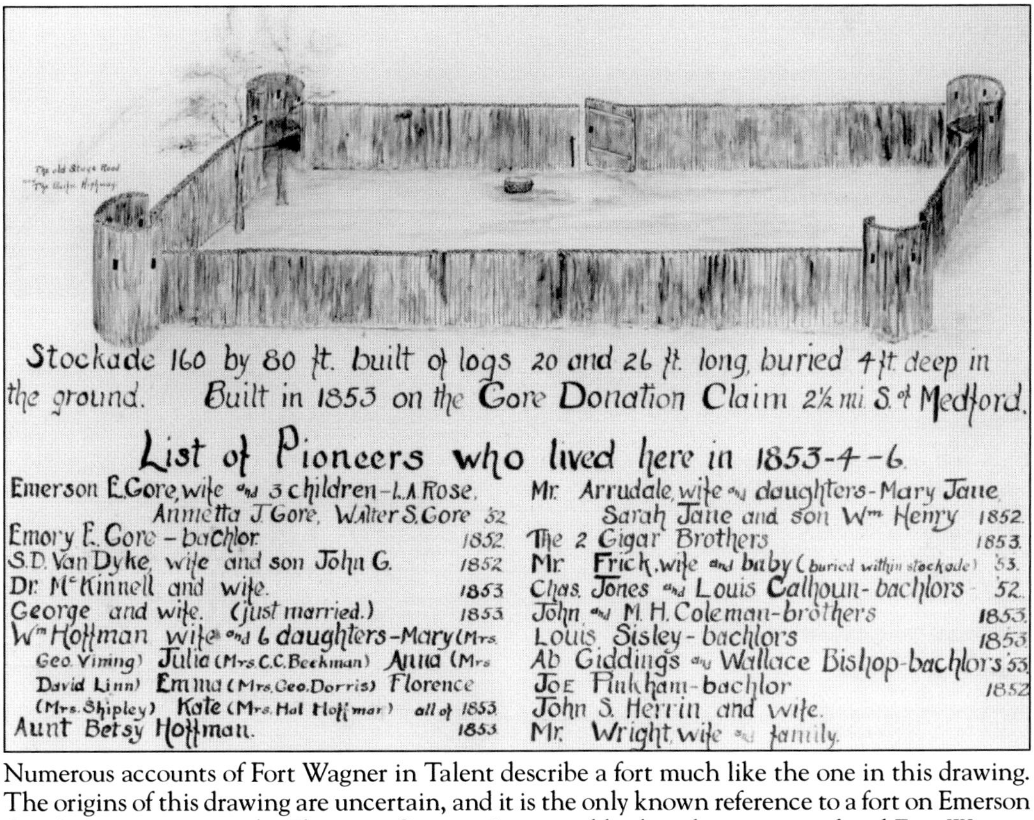

Numerous accounts of Fort Wagner in Talent describe a fort much like the one in this drawing. The origins of this drawing are uncertain, and it is the only known reference to a fort on Emerson Gore's property in nearby Phoenix, Oregon. It is possible that the artist confused Fort Wagner with the Gore place. The listed occupants of the fort were, however, pioneers of Phoenix not Talent. (SOHS No. 7956.)

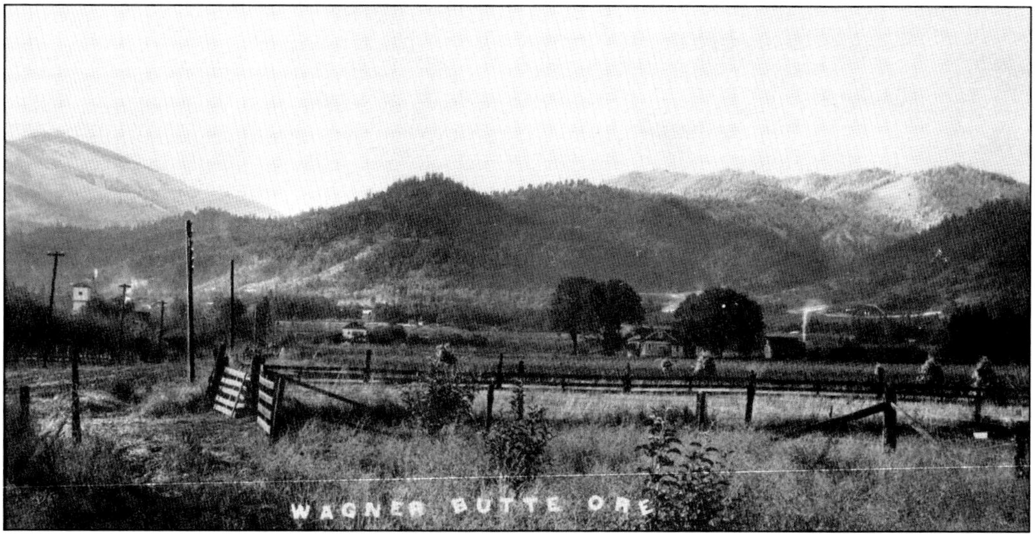

This timeless view endures and embraces the town no matter what the season or how much development takes place. The ridges between Wagner and Anderson Peaks are the backdrop to all the lives lived out in this corner of the world.

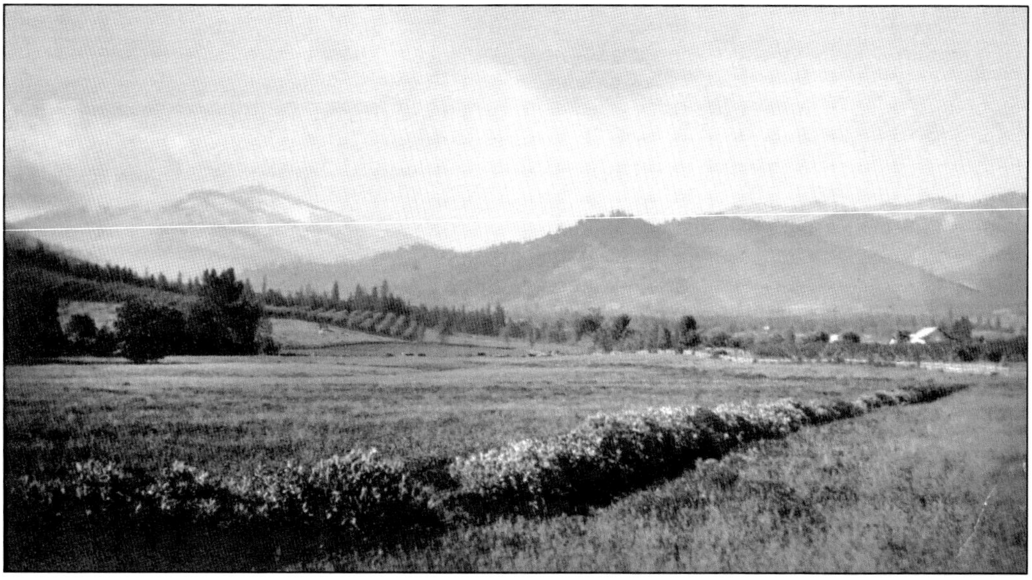

The Beeson-Robison ditch cut a swath through agricultural lands from early settlement days. Talent Irrigation District took advantage of many of the ditches in and around Talent to form a network of channels to supply water for crops all over the valley. (Harry Gleim.)

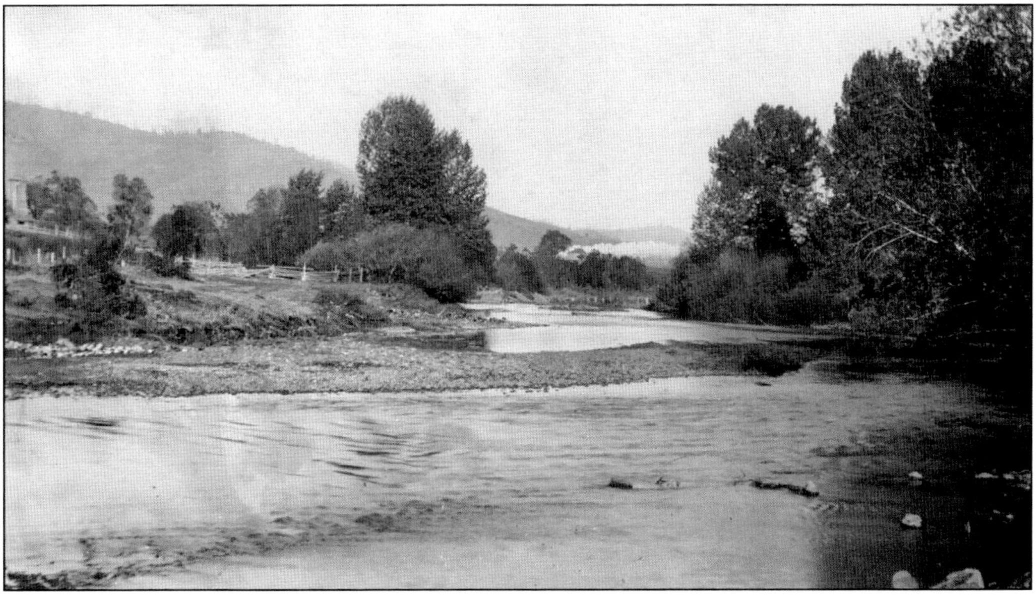

Because the road to California and the Applegate Trail followed Bear Creek through the valley, it was the center of action for explorers, trappers, Native Americans, and pioneers. Local Native Americans did not drink from it but used springs instead.

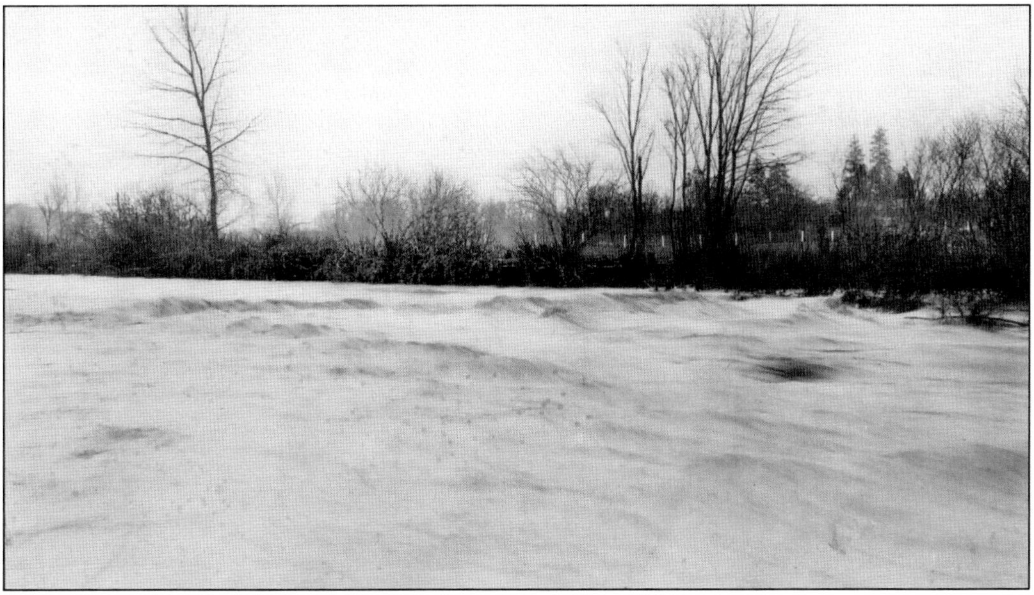

Bear Creek is shown here in 1917 running over its banks. Through the years, floods have wiped out burial grounds, bridges, homes, crops, and fences. In spite of this, development still takes place in the floodplains.

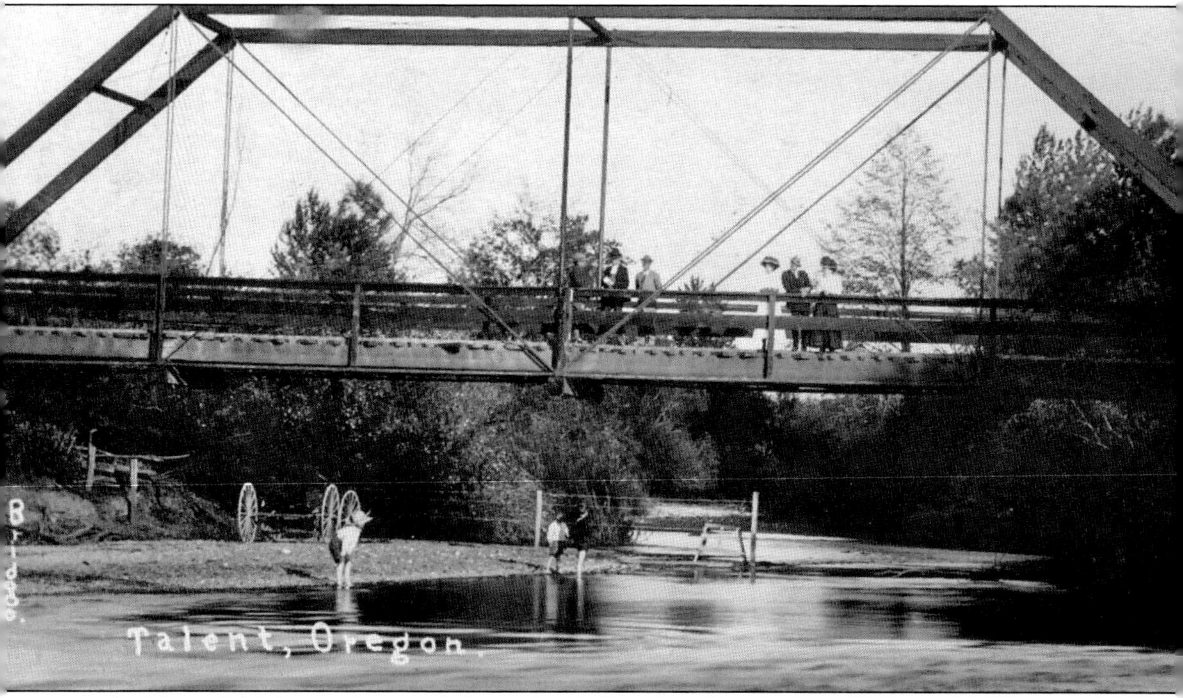

As automobiles became more common, bridges were needed to ford waterways. This bridge may be the one that once crossed over Bear Creek at Valley View Road near the present-day Wal-Mart. (Mike Cotta.)

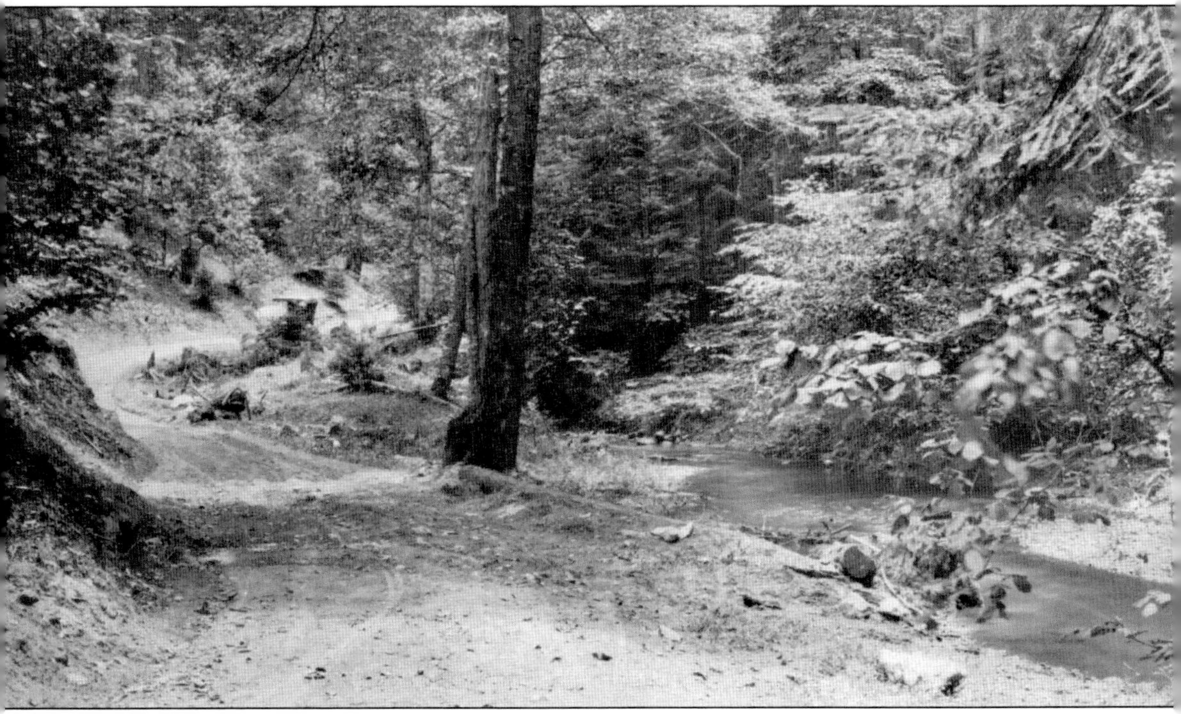

Wagner Creek Road follows the creek to important resources. Fishing, hunting, hiking, mining, logging, sawmills, and schools have all been located along this road and creek. Salmon were once abundant in all the waterways of the Rogue Valley.

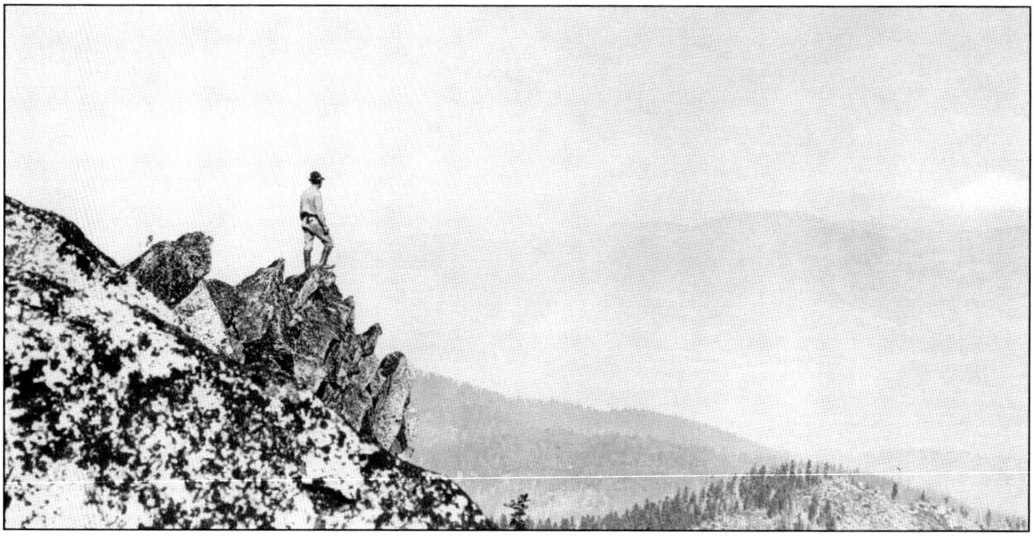

Climbing to the top of Wagner Butte from the trailhead is about a 10-mile round-trip. Charles Holdridge is seen on the rocks just below the summit and may have been one of the hikers who carved his initials in the rocks.

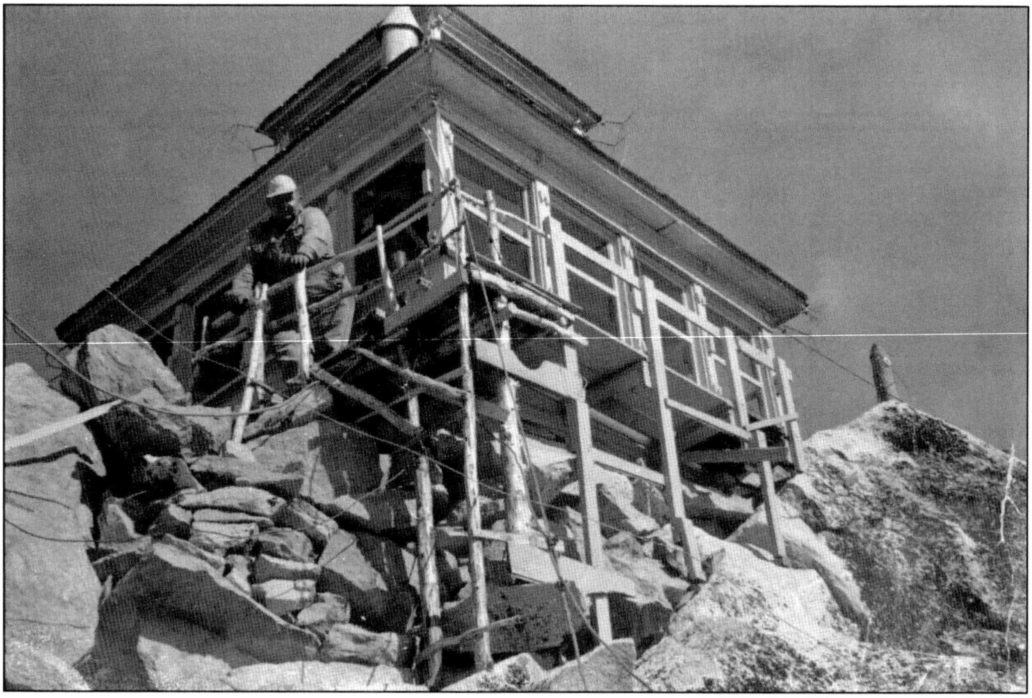

It was a coveted summer job to be the ranger at the fire lookout on top of Wagner Butte. This lookout with a cupola on top was built in 1923 and replaced in 1961. The second building was completely torn down and some of the rocks blasted away in 1971.

Lightning-caused fires from summer storms often fill the valley with smoke. This photograph documents a fire on nearby Mount Ashland as seen from the Holdridge orchard along Rapp Road about 1910.

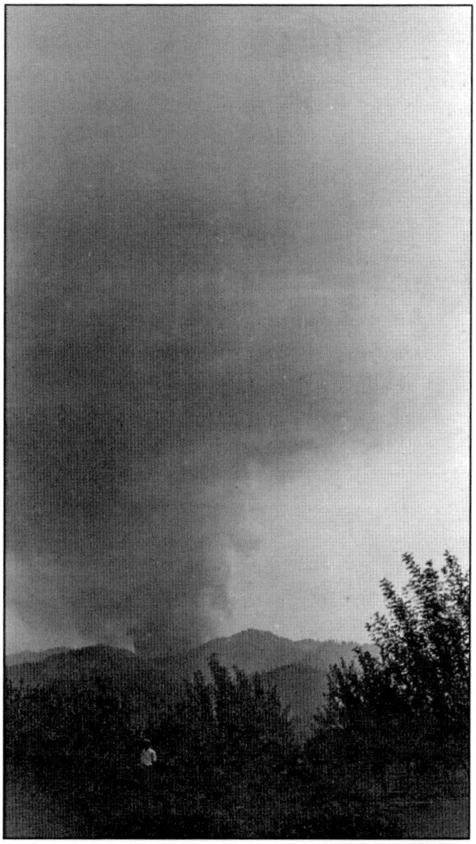

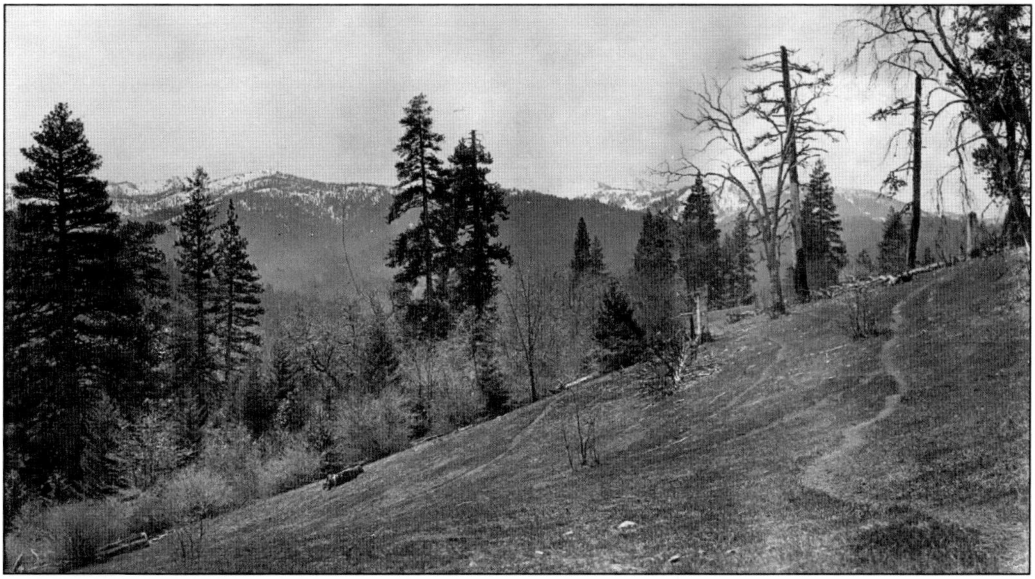

Anderson Gap reaches to the extreme western boundaries of the Talent area. It is well outside the city limits but absorbed a lot of activity from gold mining to logging in the early days. It is named after Eli K. Anderson and his brother, James Firman Anderson, who had adjoining land claims on the valley floor.

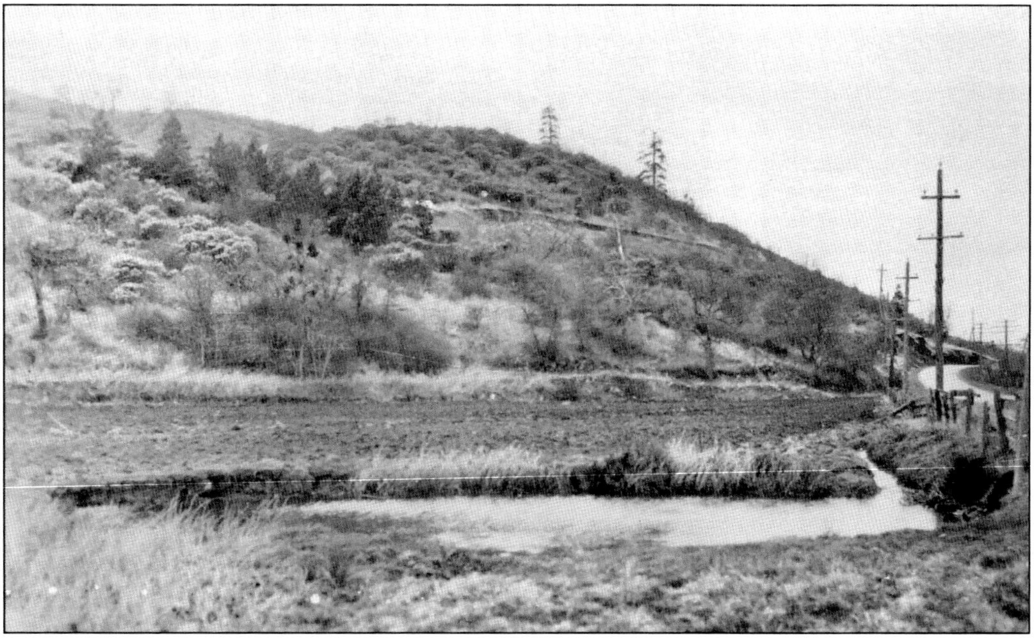

Now called Jackson Hot Springs, this warm spring was right along the road between Talent and Ashland. Long before the pioneers arrived, Native Americans used the healing waters of this spring. (Kerby family.)

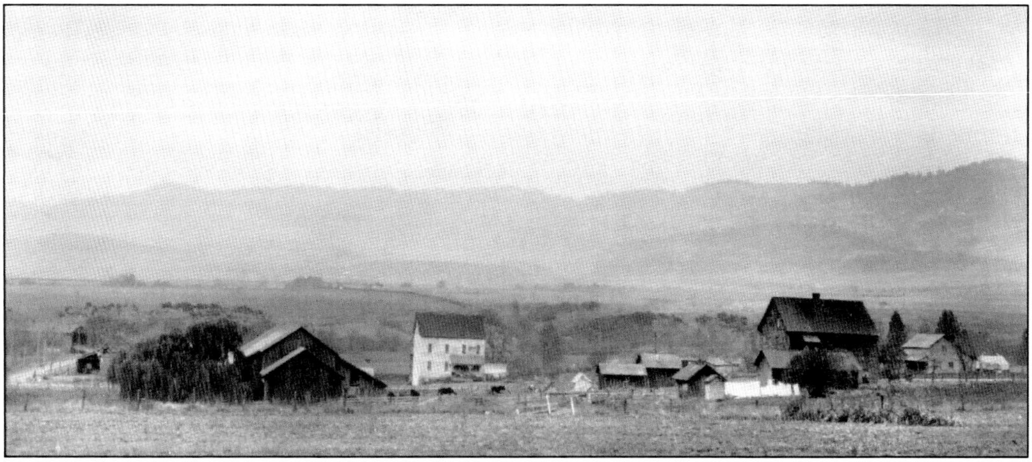

Built in 1854, Eagle Mills was the closest gristmill for Wagner Creek farmers. Remnants of some of the buildings are still standing along Jackson Road north of Ashland. The covered bridge across Bear Creek (seen on the left) is no longer there. (Kerby family.)

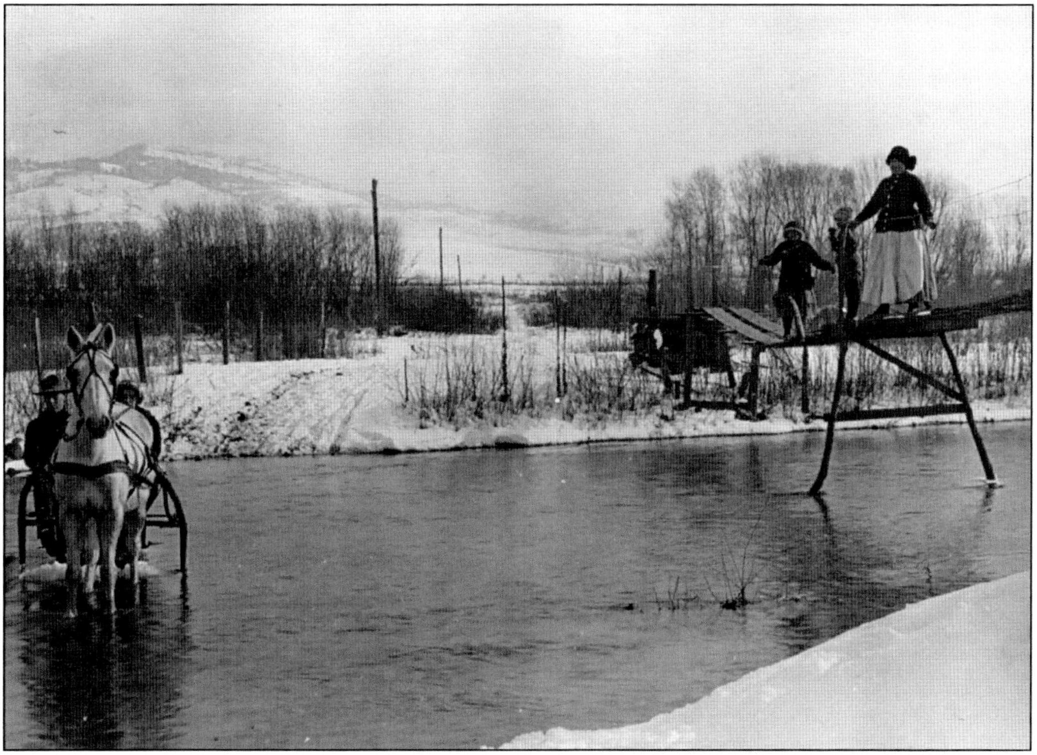

Winter cold did not stop Floyd and Angie Bell from crossing the icy waters of Bear Creek in a buggy rigged with winter runners. Holdridge family members took the walking bridge.

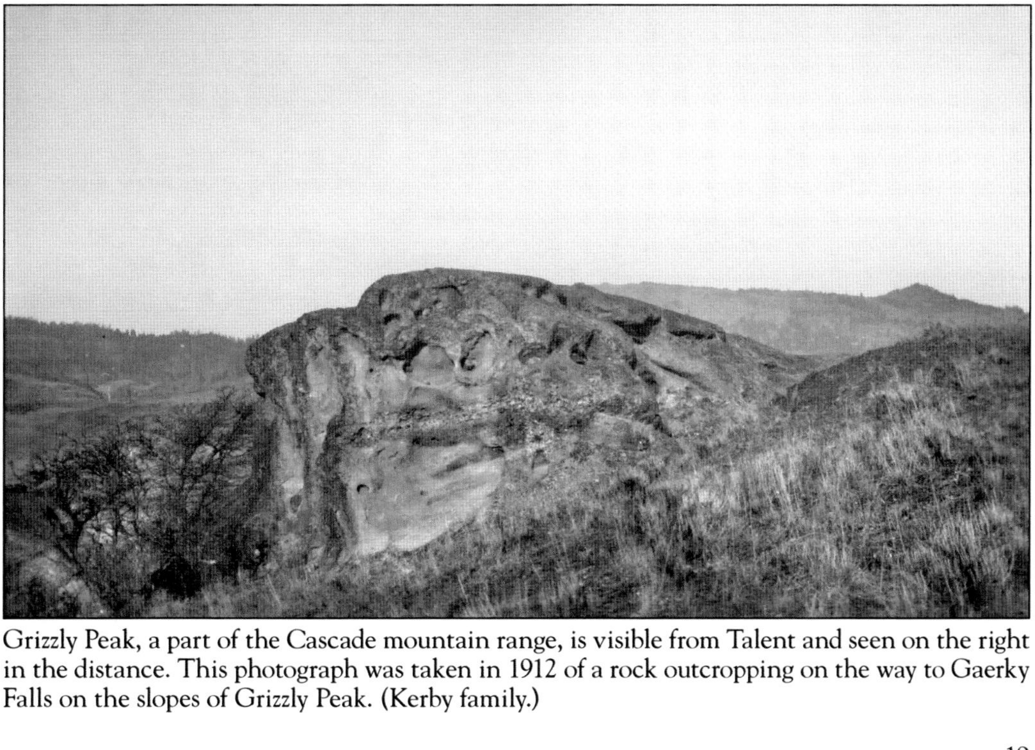

Grizzly Peak, a part of the Cascade mountain range, is visible from Talent and seen on the right in the distance. This photograph was taken in 1912 of a rock outcropping on the way to Gaerky Falls on the slopes of Grizzly Peak. (Kerby family.)

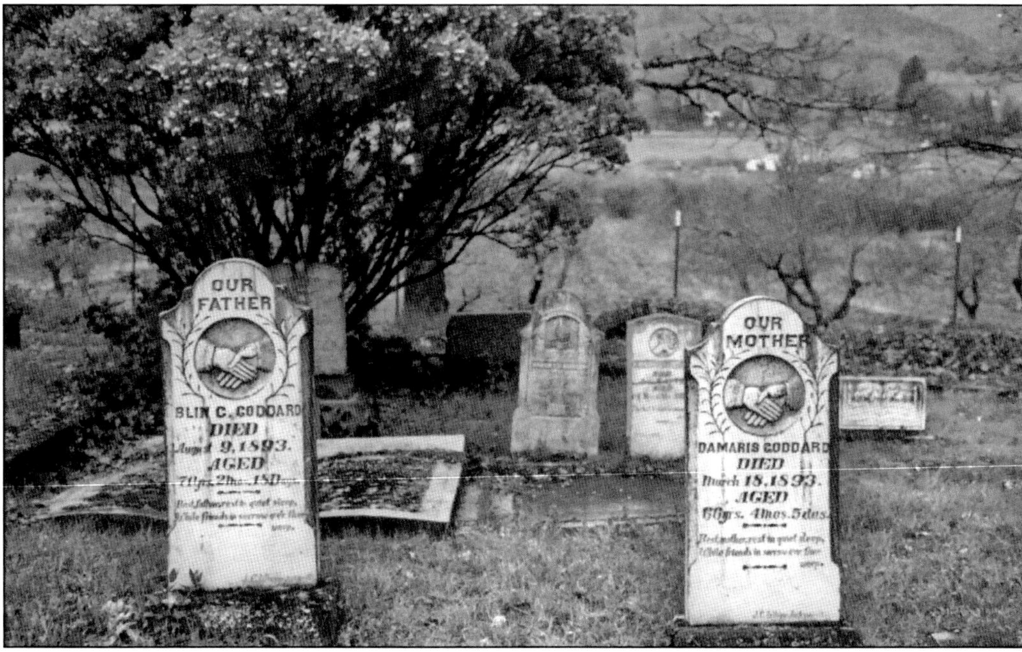

In 1857, Avery Stearns was the first to be buried in David Stearns's wheat field near the corner of Wagner Creek and Anderson Creek Roads. Though it was used as a final resting place for many pioneers, it did not become a public cemetery until David died in 1886 and the land was deeded to the Wagner Creek Cemetery Association.

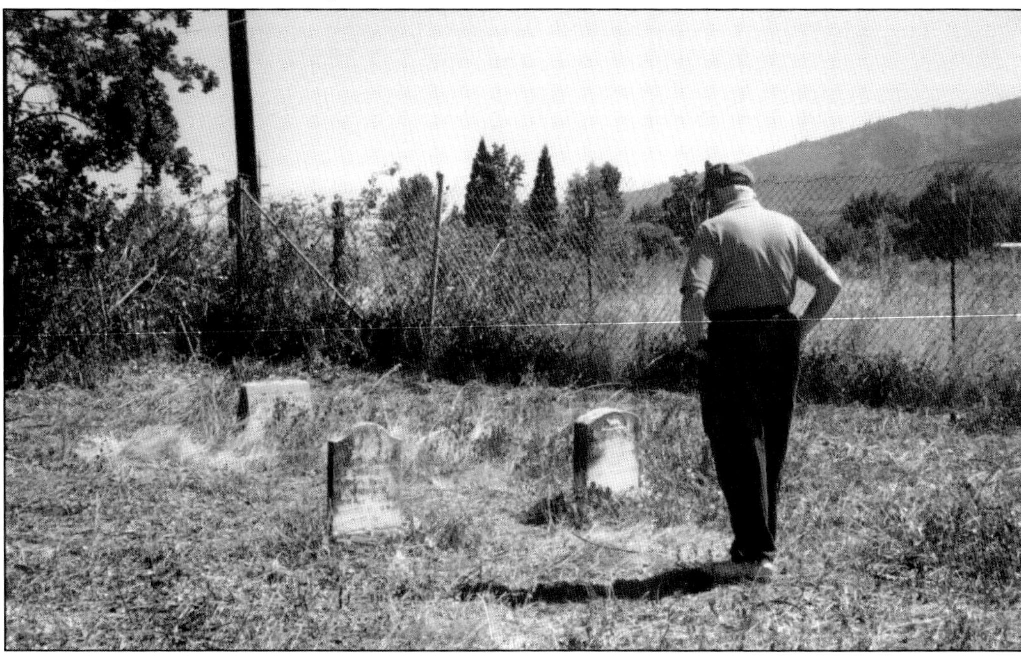

Ralph Hunkins, a Talent Historical Society board member, was photographed examining the Church of the Brethren or Dunkard Cemetery north of town. There was once a church house near the cemetery, but the church was later moved to Ashland. The Dunkards got their name from the method of baptism they practiced.

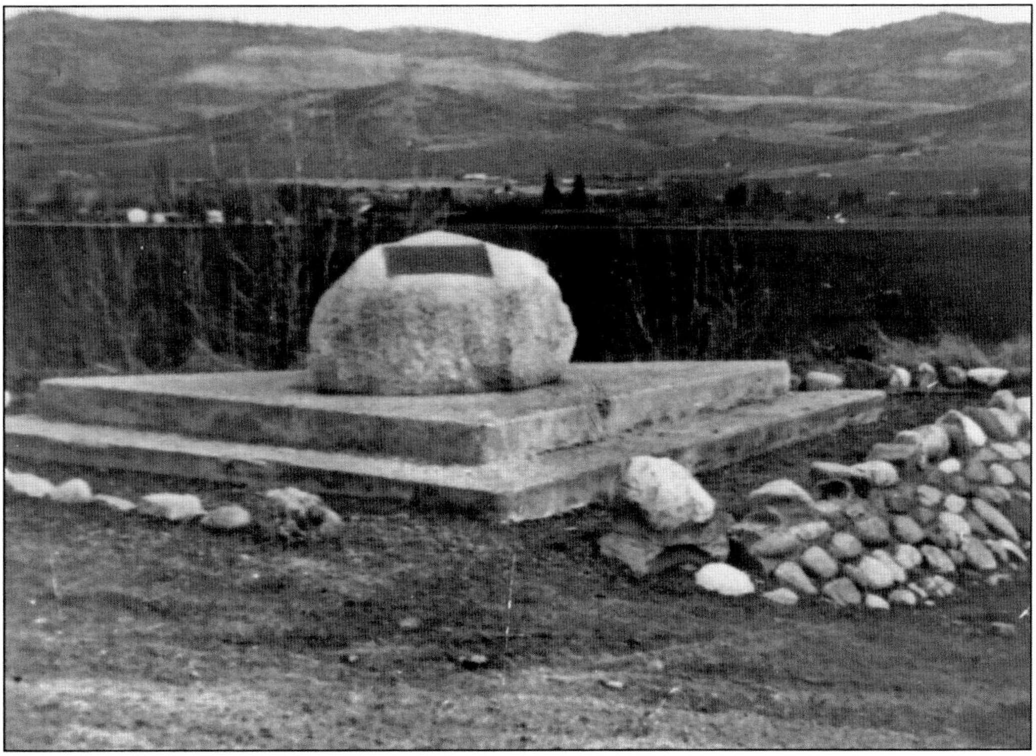

The monument located on the corner of Rapp and Wagner Creek Roads marks the approximate spot where Jacob Wagner obtained the first water rights in Oregon in 1852. The monument was erected in 1952 on an empty field, which is now covered with houses. (SOHS No. 1082.)

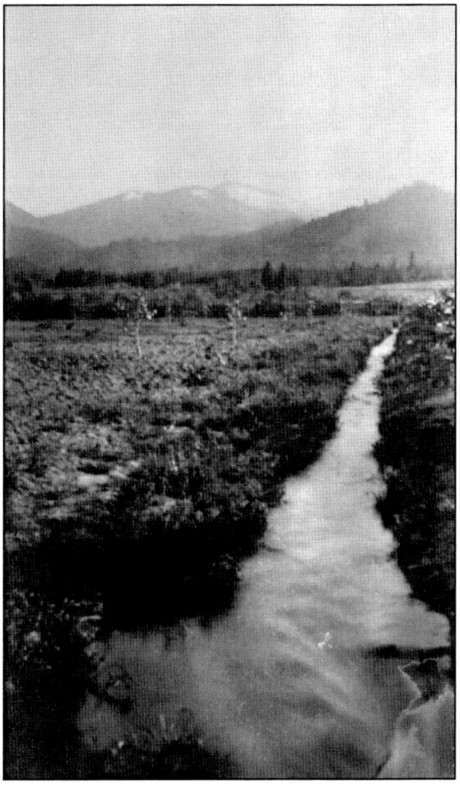

While other towns in the valley used more of their water resources for mining and for powering mills, Talent was quick to capitalize on the abundance of water for agricultural purposes. A series of ditches were dug by hand and maintained cooperatively. (Harry Gleim.)

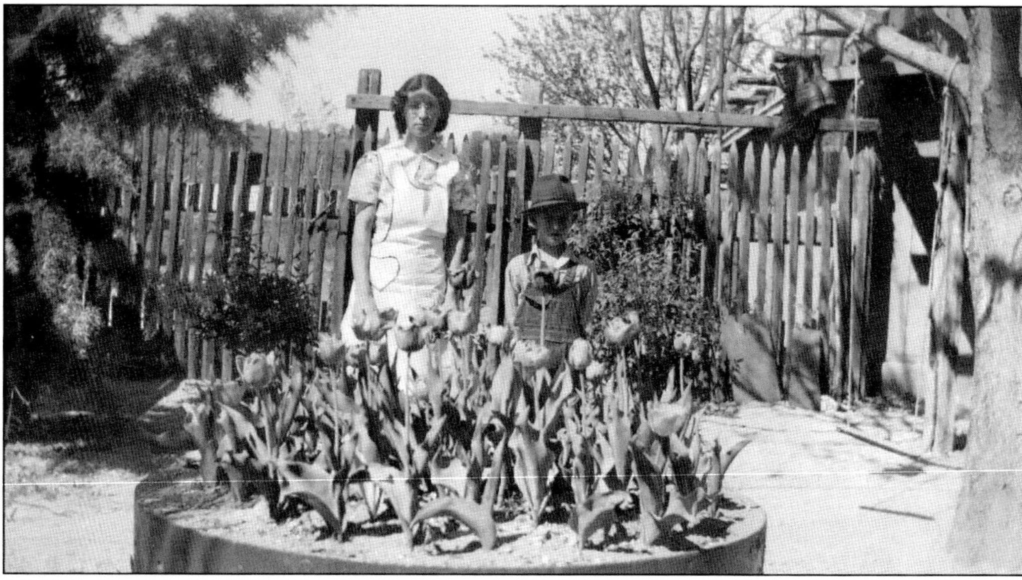

Still in her apron, Marguerite Kime Riggs stood beside the circle of tulips she carefully planted. Her husband's boots hanging in the tree to the right make the scene even more charming. Women often created their own landscape and fenced it in to protect it from animals. (Jerry Kime.)

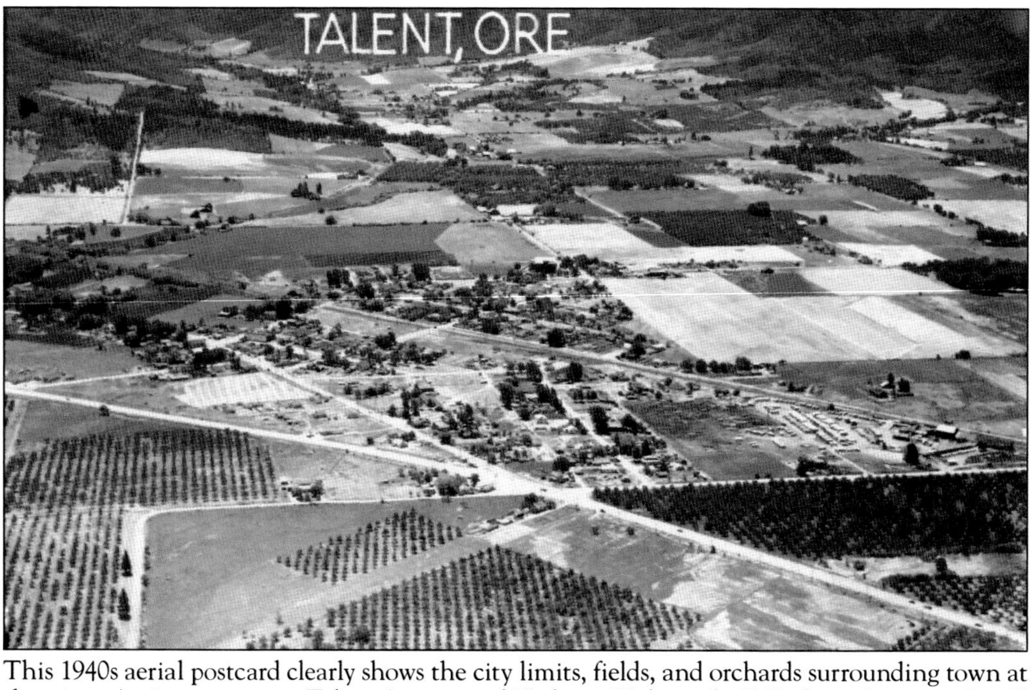

This 1940s aerial postcard clearly shows the city limits, fields, and orchards surrounding town at that time. An intersection at Talent Avenue and Highway 99 formed a "V" where they met. A mill stood on the site where the Candlewood Mobile Home Park is now located along Colver Road.

Two

PEOPLE

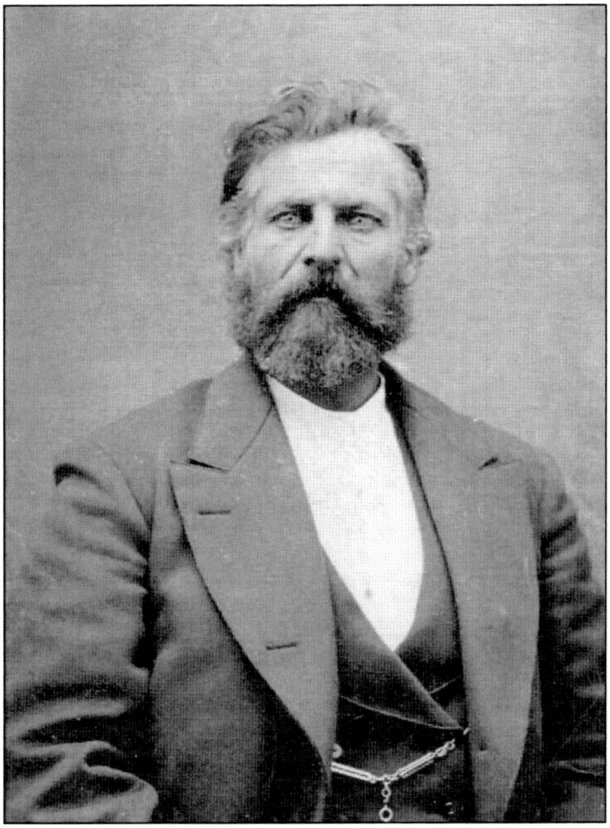

Welborn Beeson was a teenager when he arrived with his parents in southern Oregon in 1853. His dairies from 1851 to 1893 left a powerful historical record of the cultural, social, political, and personal events that touched the lives of southern Oregonians. His honest and detailed accounts distinguish his writing from others. Talent Historical Society is transcribing his entire diary, making it available to the public. (Carter family.)

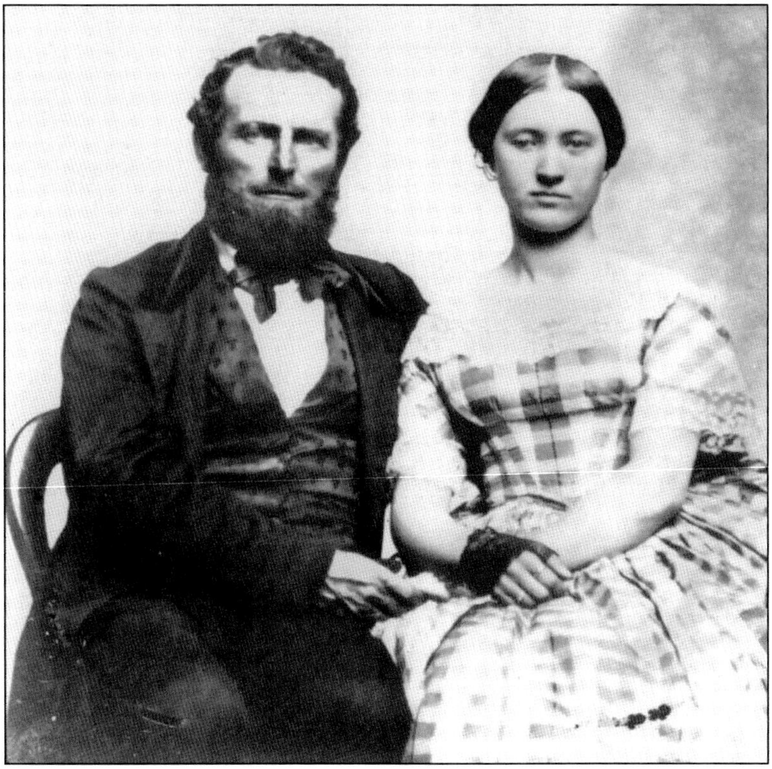

Jacob Wagner was a bachelor when he came to settle on Wagner Creek in 1852. In 1860, he traveled back to Iowa to marry 19-year-old Ellen Hendrix and bring her to Oregon. Jacob was more a businessman than a farmer, and in 1861, the couple moved to Ashland, Oregon, where he invested in the flour mill and other enterprises. (SOHS No. 1384.)

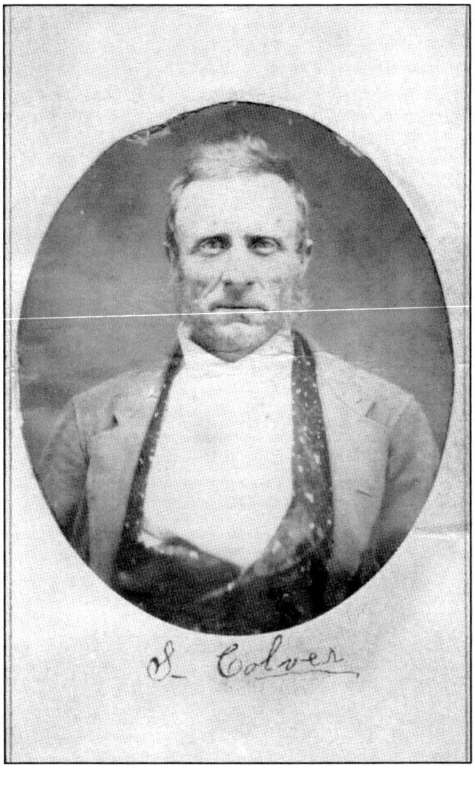

Samuel Colver served briefly as an Indian agent at Fort Lane near Table Rocks. His family home stood in Phoenix, Oregon, until 2008, when it was damaged by fire and torn down. Samuel died in 1891 when his horse broke through an icy lake in Klamath County. It was months before they found his body. (Maxine Kincaid.)

Abolitionist John Beeson came from Illinois to Oregon in 1853. When he moved to the West, his focus changed from slavery to the rights of Native Americans. As a result of his outspokenness, he was pressured to leave his family and the valley in 1856. He went back east and published *A Plea for the Indians* in 1857 and for the rest of his life advocated for civil rights for the Native Americans, speaking to churches, politicians, presidents, and Indian agents. (Carter family.)

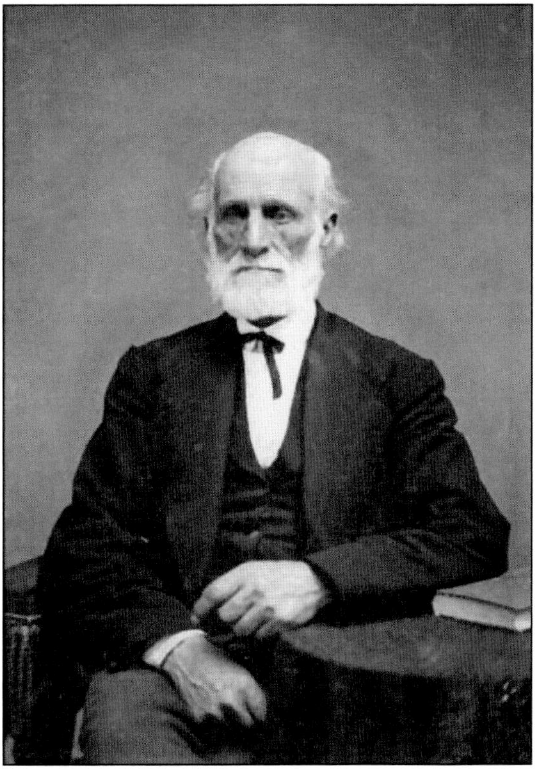

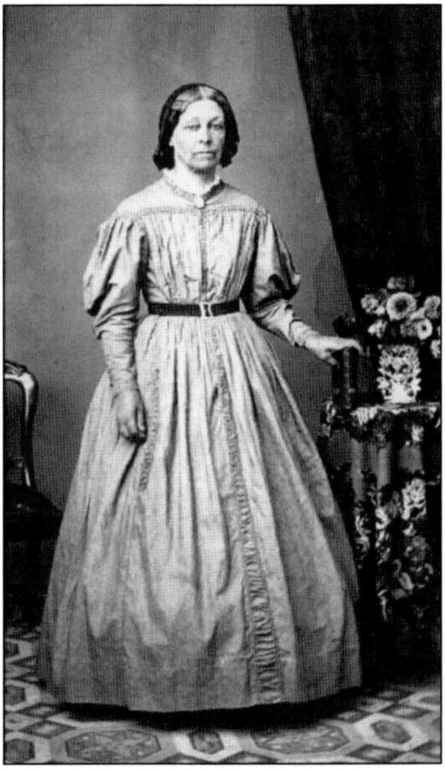

The wife of John Beeson and the mother of Welborn, Ann Beeson had to struggle through pioneer times without her husband's help. In fact, he often asked his family for money for his mission for the Native Americans in the eastern states. Ann stood for this portrait at the Peter Britt Studio in Jacksonville, Oregon, in 1864, two years before she died. (Carter family.)

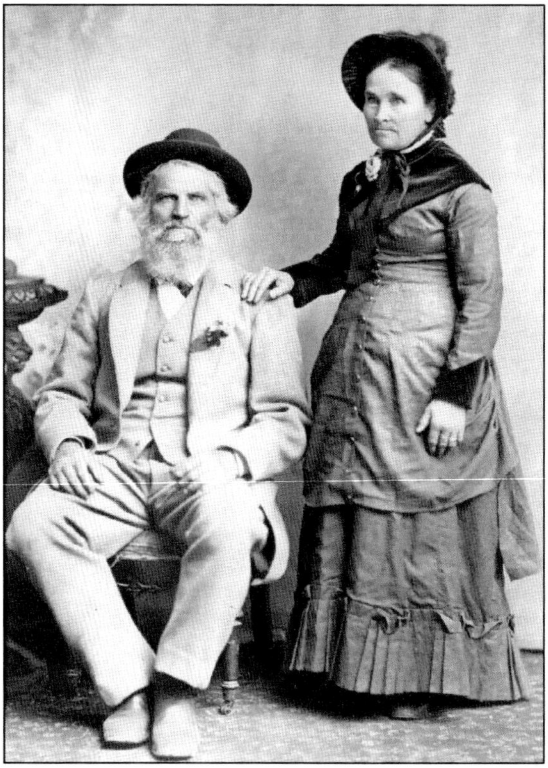

This adorable couple was Isaac and Catherine Wagner. Isaac, the brother of Jacob Wagner, moved to Oregon in the early 1860s to take over Jacob's farm. The family added improvements to the farm and was able to move out of the old Fort Wagner log cabin when they built their own frame house. (Carter family.)

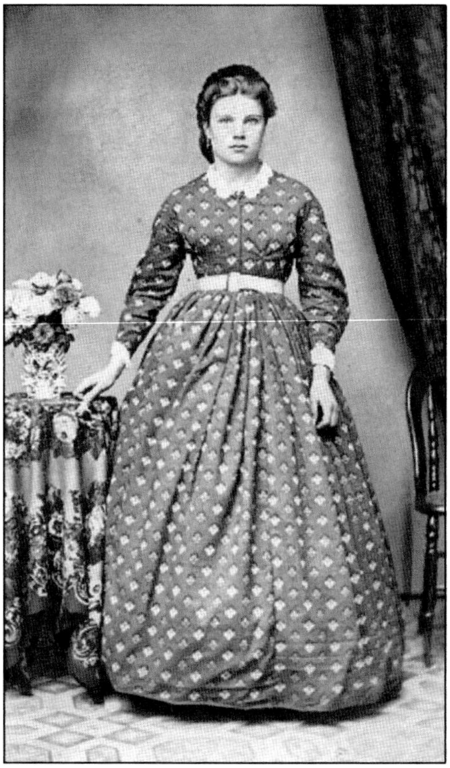

Esther Ann Wagner, daughter of Isaac and Catherine Wagner, was a local beauty. Many eligible bachelors, including Welborn Beeson, sought her attention. He would have married her if she had accepted his proposal. This Peter Britt portrait was taken in 1864. (Carter family.)

Company I of the Oregon Calvary was stationed at Fort Klamath at the time this photograph was taken in the 1860s. Ira Chandler (seated left), Julius Foss (standing right), and Lewellyn Colver (seated right) were local men who joined the U.S. Calvary during the Civil War. The men in the photograph were friends of Welborn Beeson, and they often wrote letters to him about their military adventures. (Carter family.)

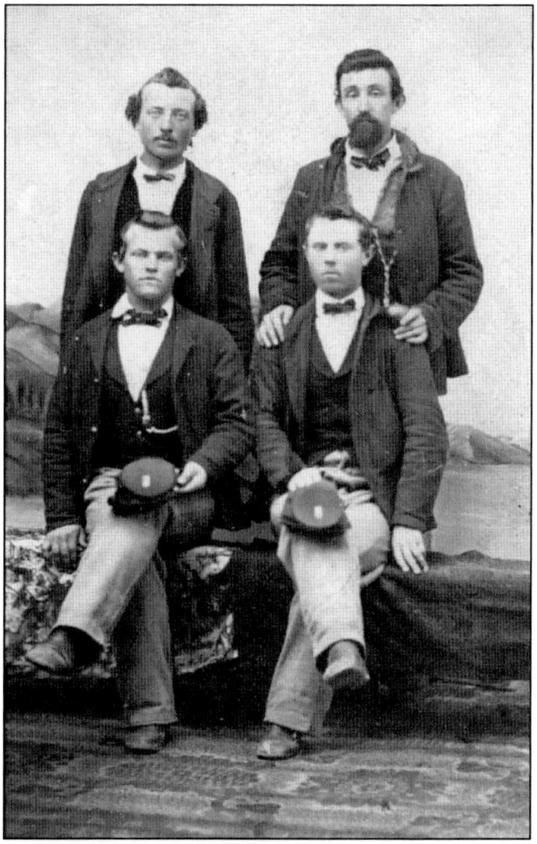

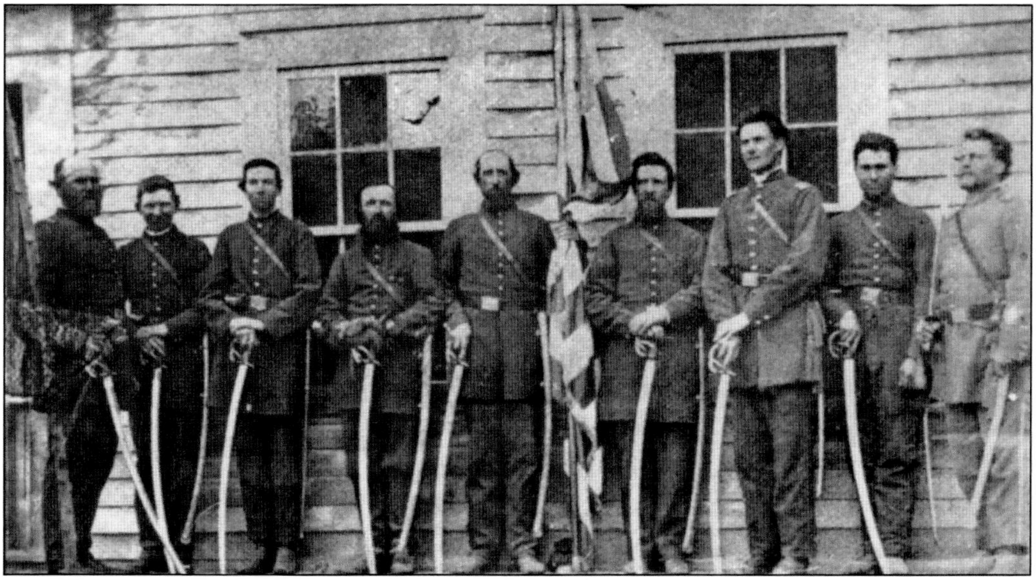

The Mountain Rangers served as the home guard during the Civil War. They fought no battles but drilled and marched and had plenty of dances. From left to right are Abel Helman, F. W. Myers, O. C. Applegate, Thomas Osborne, Minus Walker, William B. Stout, Ivan Applegate, Wallace Baldwin, and Welborn Beeson. (Terry Skibby.)

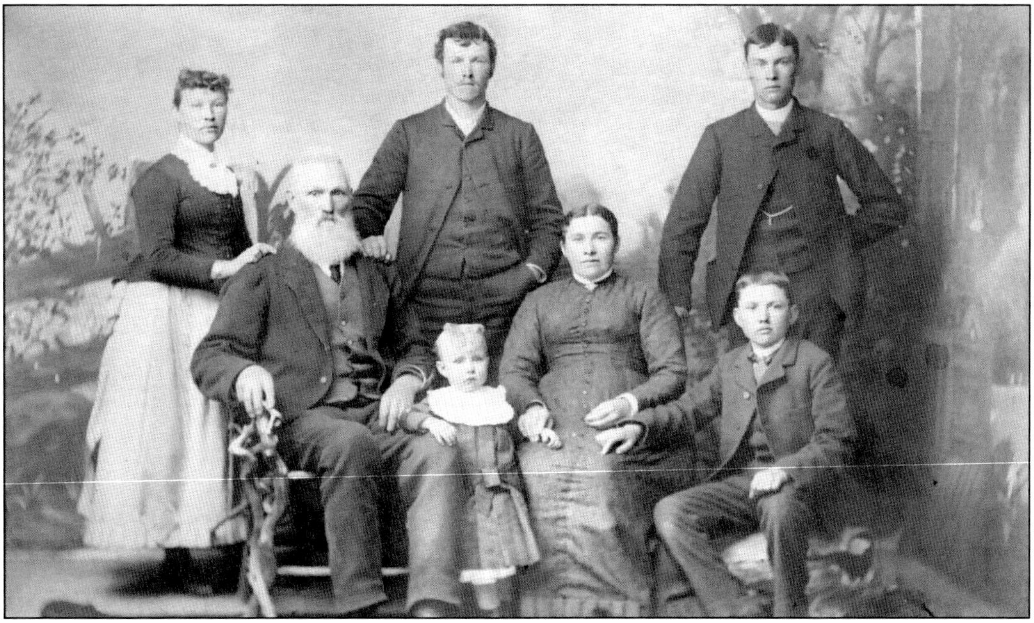

Matthew Hubbard and Sabra (Goddard) Coleman and their family had their picture taken about 1888. Matthew and his brother, John, were among the few who purchased their land from a Native American. Ironically, the woman they purchased from was a Florida native who had gotten the land from the government for service she performed against another tribe. Coleman Creek is named for Matthew and John Coleman.

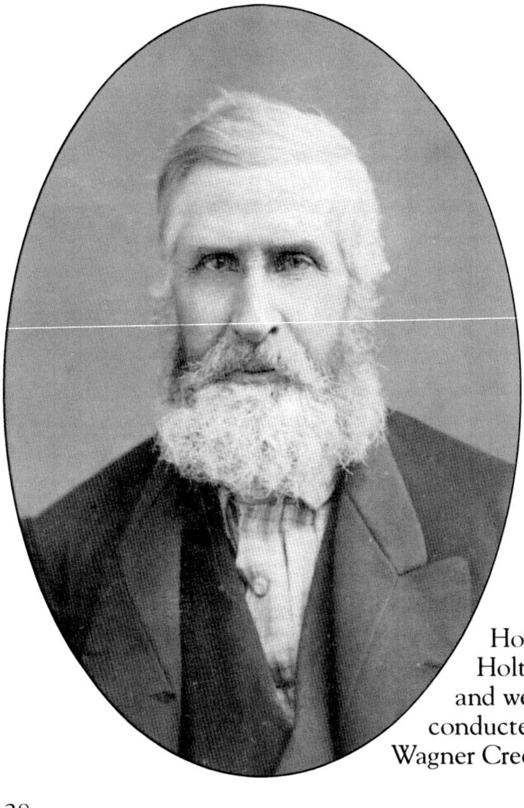

Holton Road was named for John and Hannah Holton. The Holtons were deep into spiritualism and were friends of John Beeson. Many séances were conducted under their roof to enlighten and comfort Wagner Creek residents. (Maxine Kincaid.)

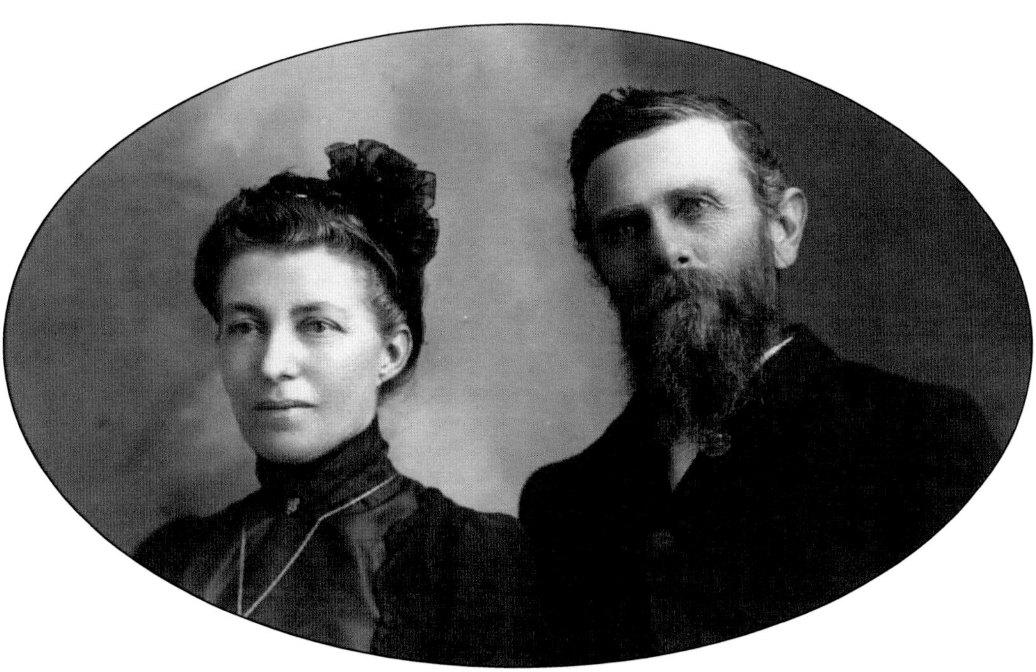

Ursula and Willis J. Dean were outspoken members of the community. Both supported freethinking and women's rights. Willis came here seeking a position as a schoolteacher. Ursula was a pretty widow with two children when they married. Copies of the Dean diary (1912–1919) can be viewed online or at Talent Historical Society Museum. (Ben Truwe.)

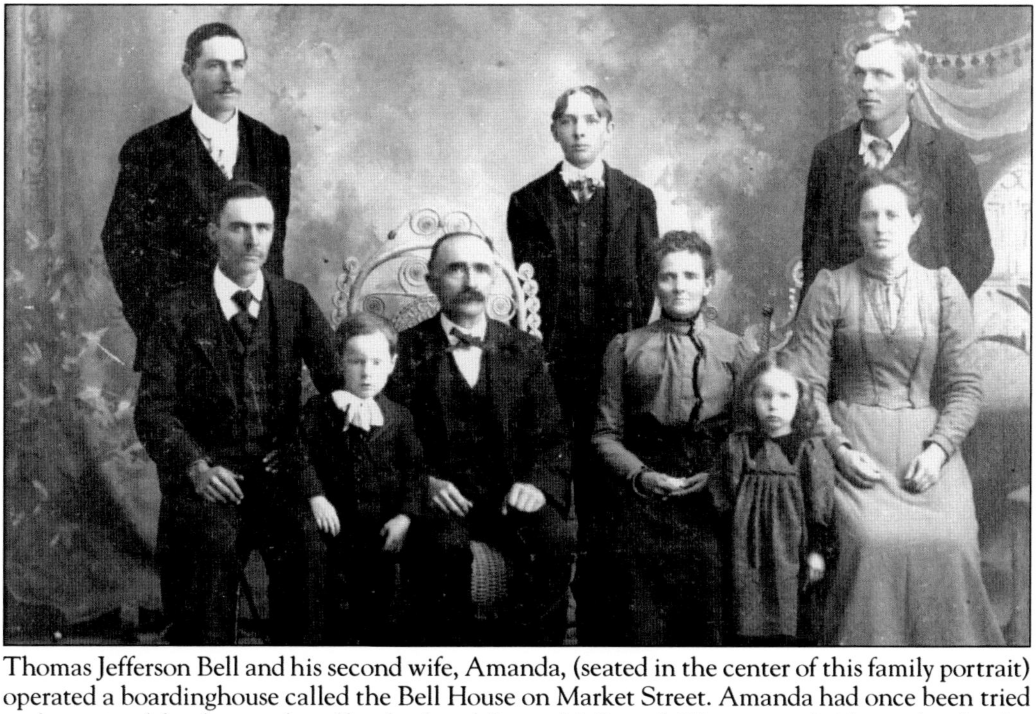

Thomas Jefferson Bell and his second wife, Amanda, (seated in the center of this family portrait) operated a boardinghouse called the Bell House on Market Street. Amanda had once been tried and acquitted for the murder of her husband, Lewis McDaniel. Thomas probably saw her for the first time when he attended the trial.

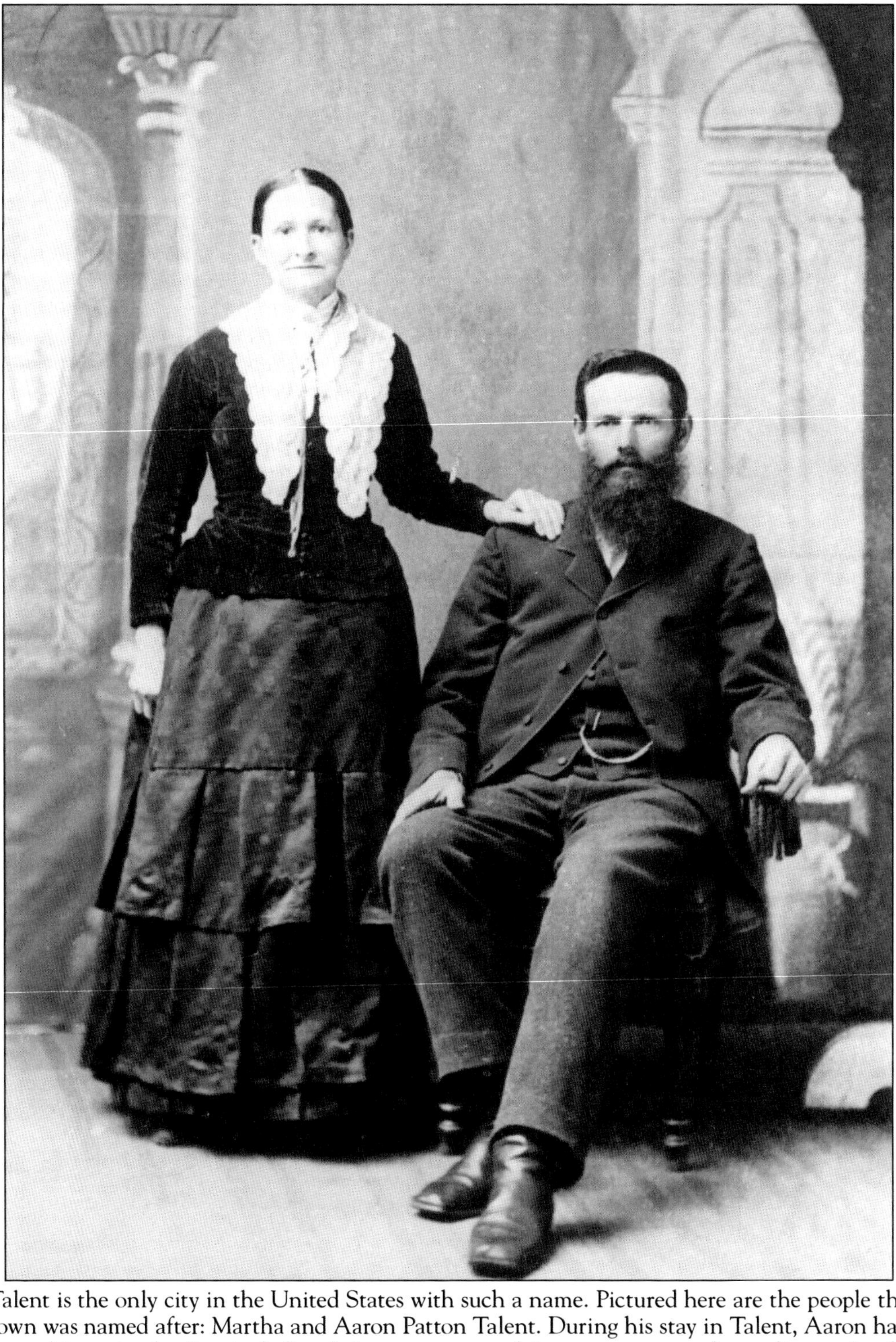

Talent is the only city in the United States with such a name. Pictured here are the people the town was named after: Martha and Aaron Patton Talent. During his stay in Talent, Aaron had several irons in the fire, including being postmaster, doing carpentry work, tending orchards, selling real estate, and operating a general store. (SOHS No. 13804.)

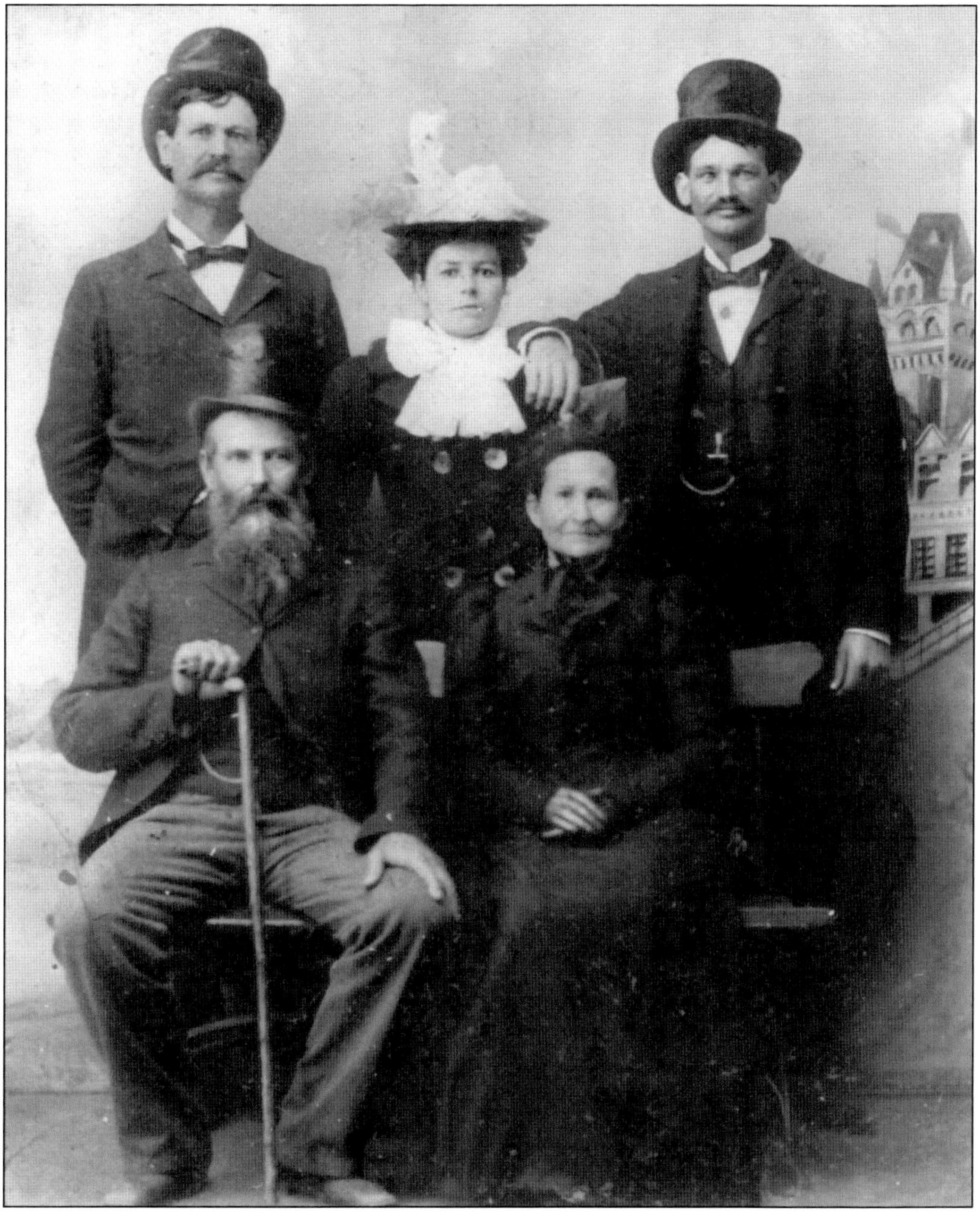

Martha and Aaron Patton Talent took the town out of the pioneer mode and propelled it into the railroad era. They are shown here with three of their children: John, Minnie, and Oscar Talent.

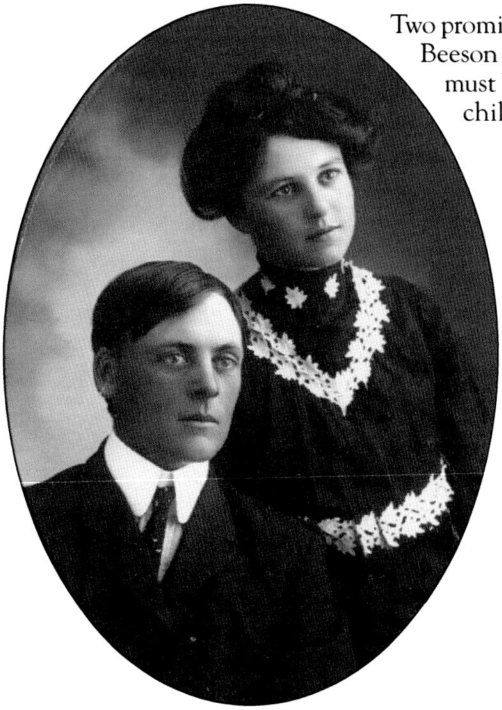

Two prominent families allied themselves when John Dalton Beeson and Myrtle Martha Talent married in 1902. They must have had many colorful stories to share with their children about growing up in Talent.

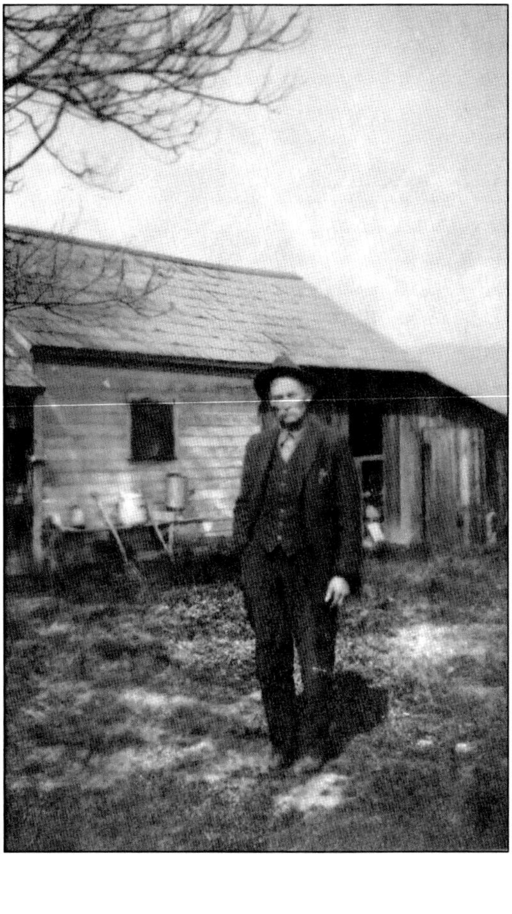

Hendrick Goddard was from a long line of carpenters. His father, Blin C. Goddard, had a hand in building many of the homes in the Wagner Creek area. A cradle made for Hendrick's firstborn in 1884 is on display at the Talent Historical Society Museum.

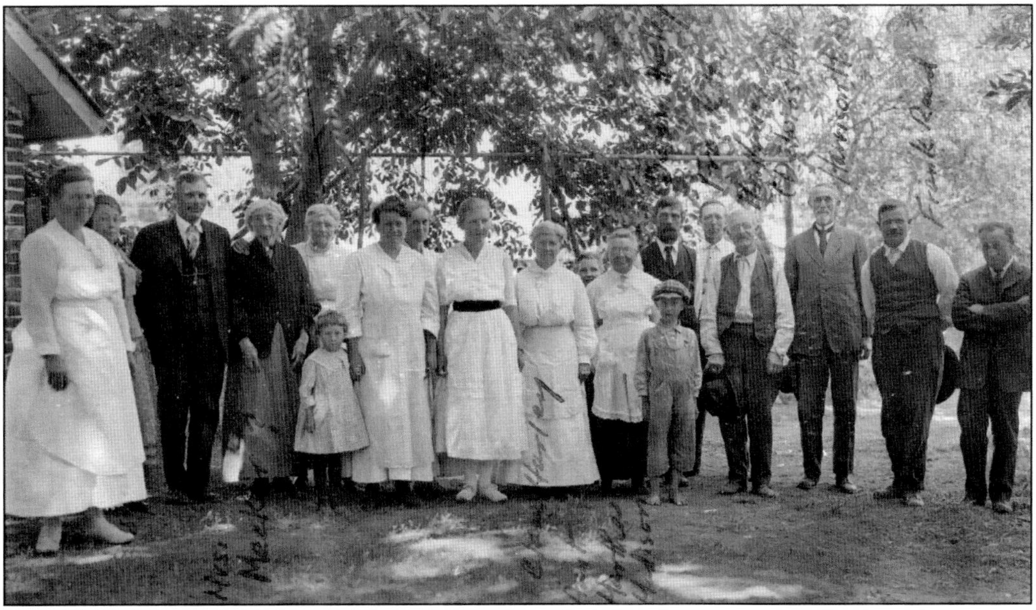

The Fourth of July was a perfect occasion for the Holdridge patchwork of relatives to get together. Aunts, uncles, cousins, and in-laws all moved from the Midwest about 1908 to Oregon and spread out all over the creek to occupy already established farms, residences, and orchards.

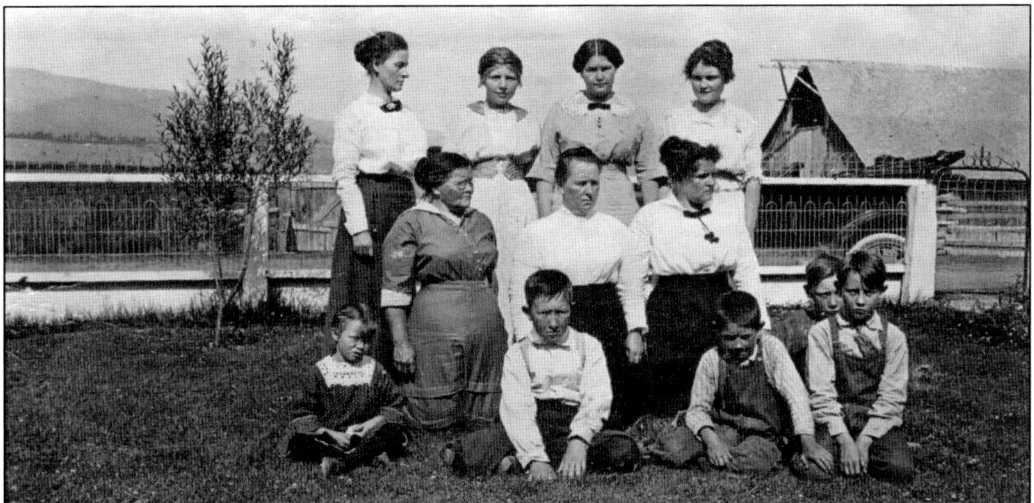

This photograph was surely taken on Sunday after church and before the dinner bell. It was produced as a postcard and was undoubtedly sent to Iowa grandparents. It is hard to tell if the Holdridge boys look grumpy because they are hungry or because they wanted to take off their ties.

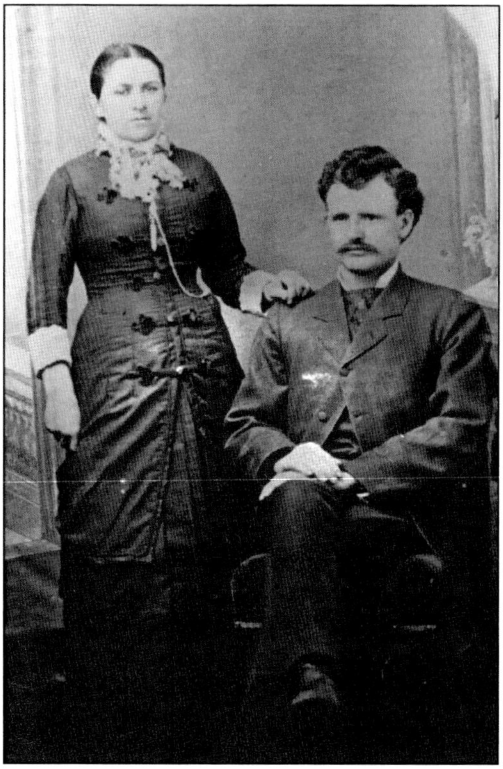

Martin and Isabella (Helms) Pellett had real estate and orchards in the Talent area. The town did not offer all the opportunities they wanted, so they moved on to California, and their orchards were eventually associated with the Newbry family. (Jerald McGrew.)

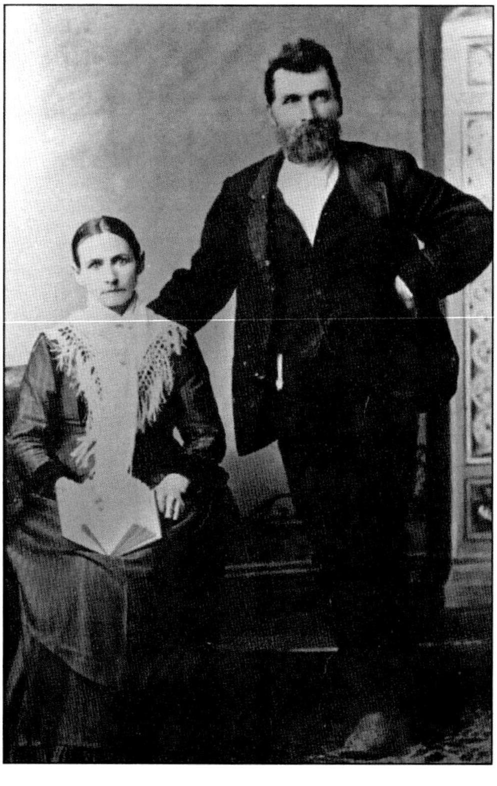

James and Elizabeth (Phipps) Helms came to Jackson County in 1863 where they made a living in cattle, real estate, and other enterprises. When they left the valley for California, their property along Talent Avenue paid off their much disputed debts. (Jerald McGrew.)

The William and Emma Kerby family lived on Yank Gulch. The name Yank is shortened from Yankee, but it is not known if it was named after anyone specific. The home where the Kerbys farmed and raised their children is gone now. (Kerby family.)

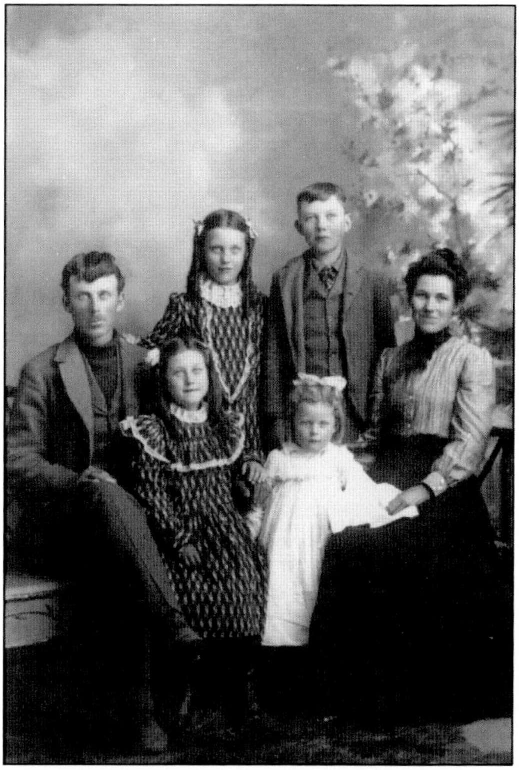

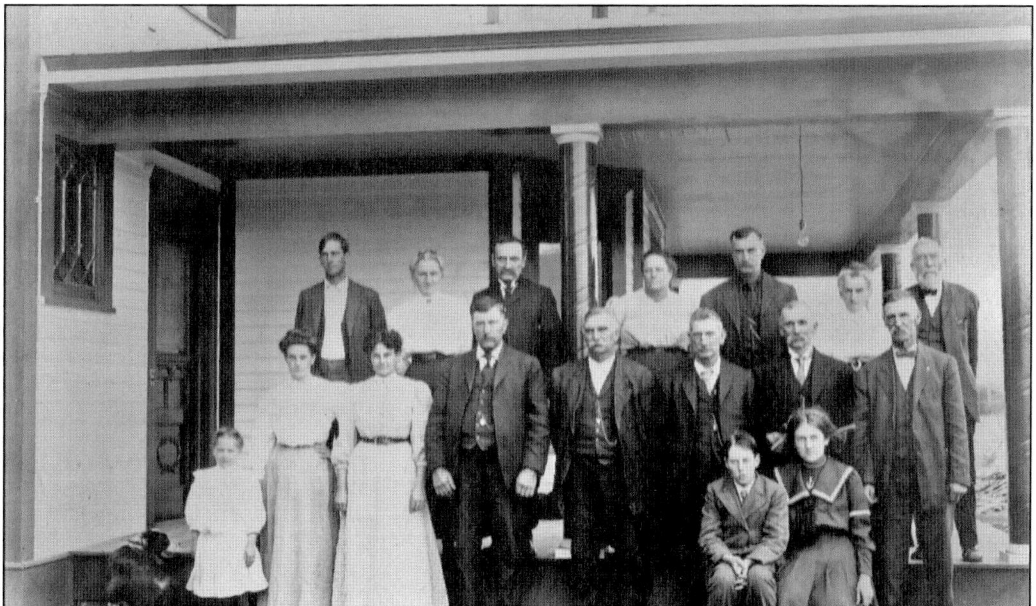

In 1909, the children of Joshua and Catherine Patterson gathered at the Patterson house just north of town. The Pattersons were prosperous enough to invest in machinery, which was used by all the farmers in the area. They charged a fee for the use of the thrasher and the road grader, which helped them pay for the equipment. (Richard Allen.)

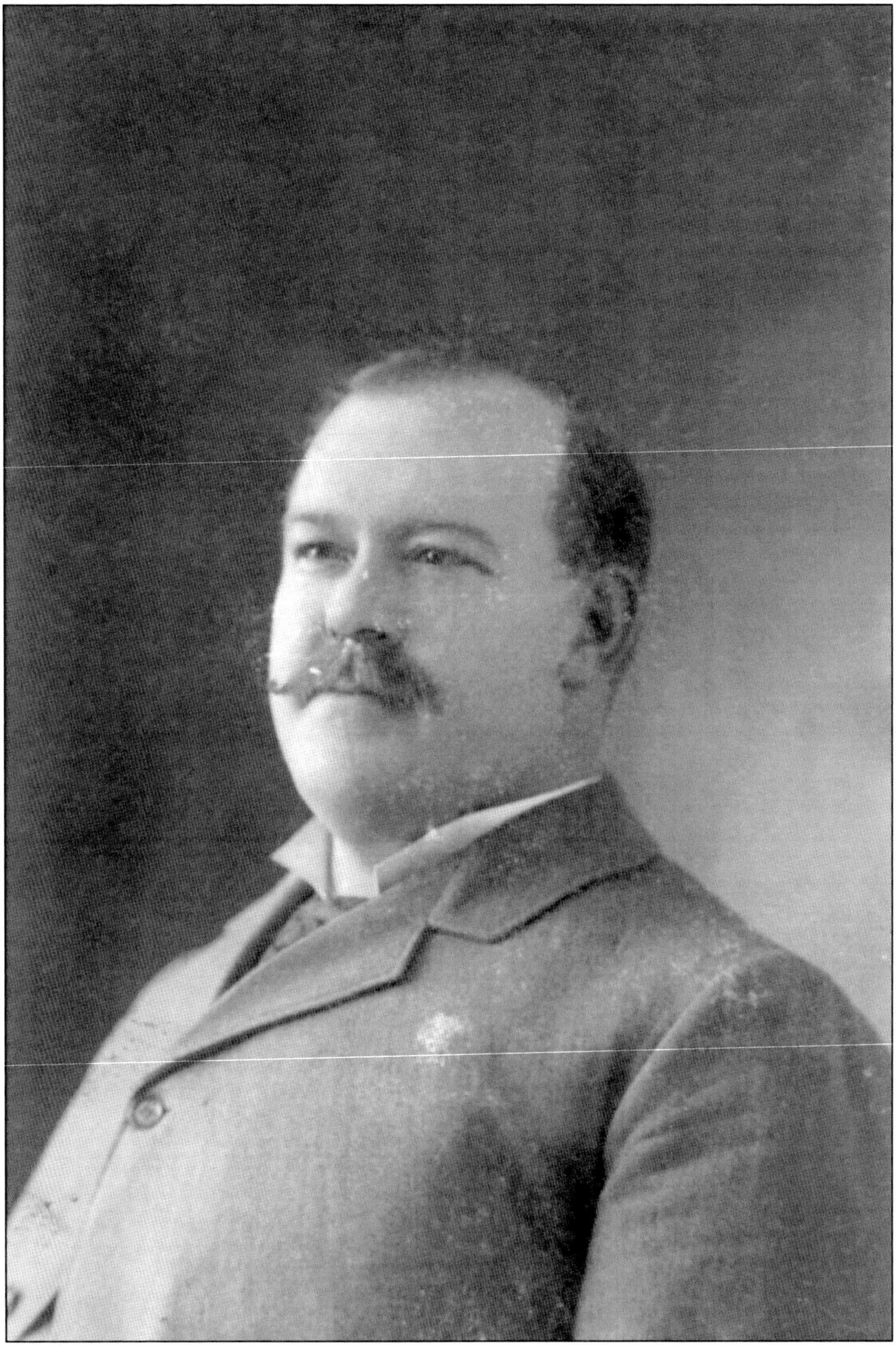

Charles W. Wolters was a big fish in a small pond in Talent. Like Ron and Sue Ridgway of modern-day Tarks Market, he was a community leader. He contributed to the growth of the town with his financial support and belief in progress. His untimely death in 1912 left his widow, Amanda Olive Wolters, and two children, Eva Pellett and Chester Wolters, to run the business. (Jerald McGrew.)

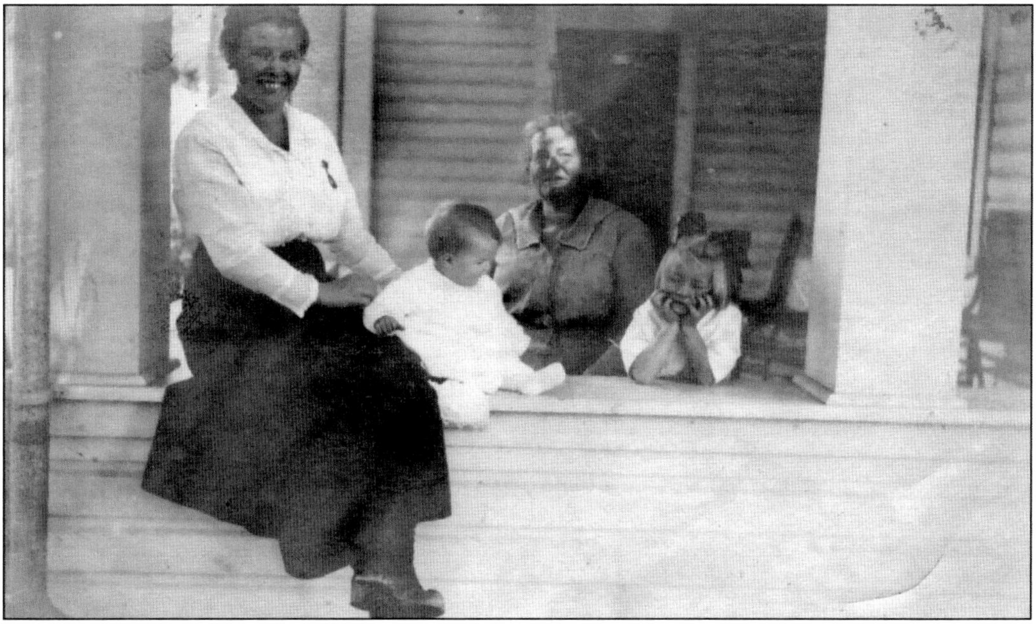

Eva Pellett and her two children had their picture taken on the porch at Grandma Amanda Olive Wolters's house on Wagner Street. The Pelletts lived next door, which made a daily trip to Grandma's an easy thing to do. Both the Pellett's and Wolters's houses still stand today. (Jerald McGrew.)

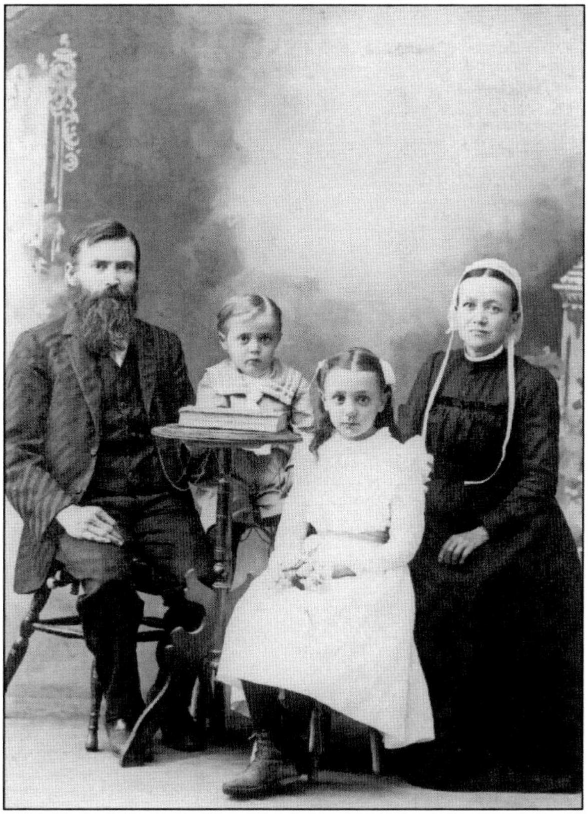

Lawyel and Anna (Springer) Settles were members of the Church of the Brethren or Dunkards. Their distinctive clothing and conservative manners set them apart from other Talent residents. They were both buried in the Dunkard Cemetery north of town. (Ralph Wyant.)

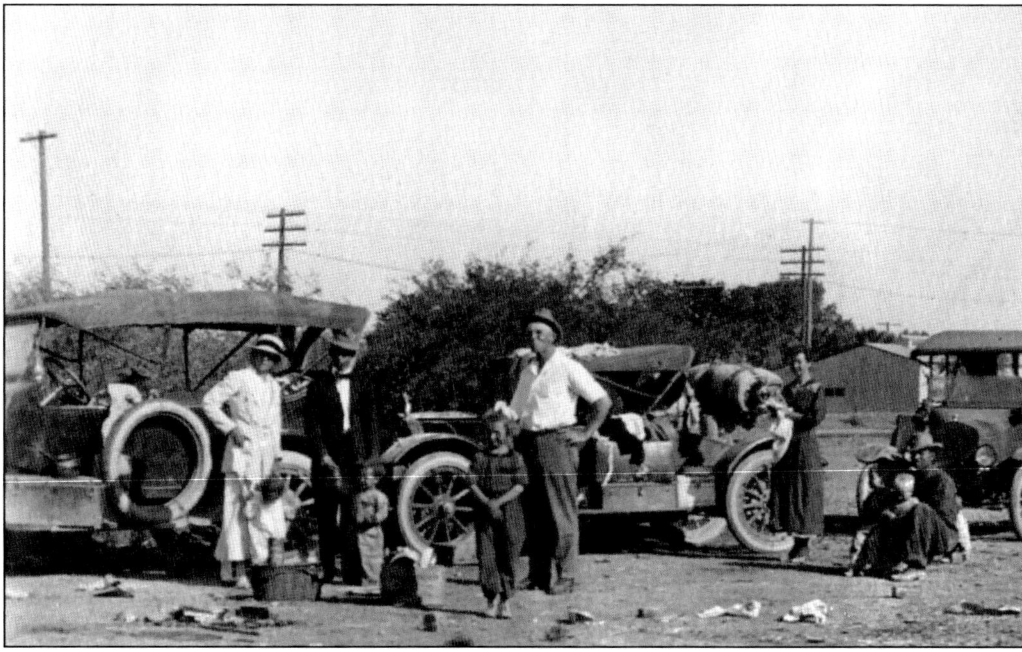

The Gleim family caravan made its final trip from El Paso, Texas, to Talent in 1918. A sleeping bag in the middle car shows that they had to camp out along the way. Young children in diapers must have been especially challenging. The man on the right with two children in his arms was Charles Will Gleim Sr. (Harry Gleim.)

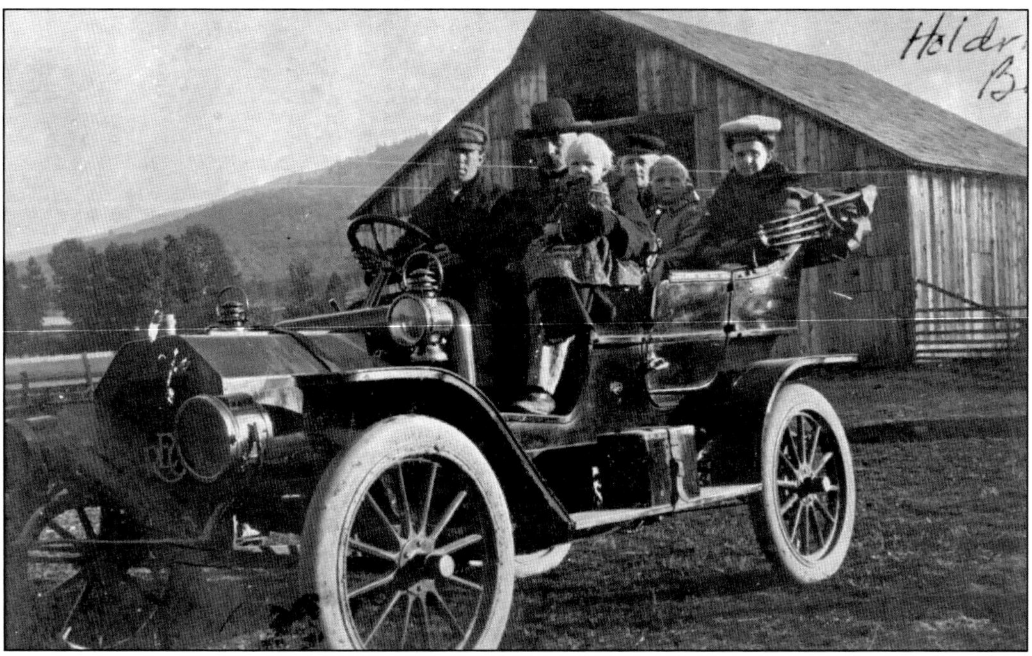

The Holdridge family bought this brand-new 1908 Reo after they came to Oregon. Driving the car with the steering wheel on the right was Frank Holdridge. Frank's father, Windsor Holdridge, sat in the front with the two women in the back, Ella (left) and Edith Holdridge. The kids filled in on anybody's lap.

Three

Dwellings

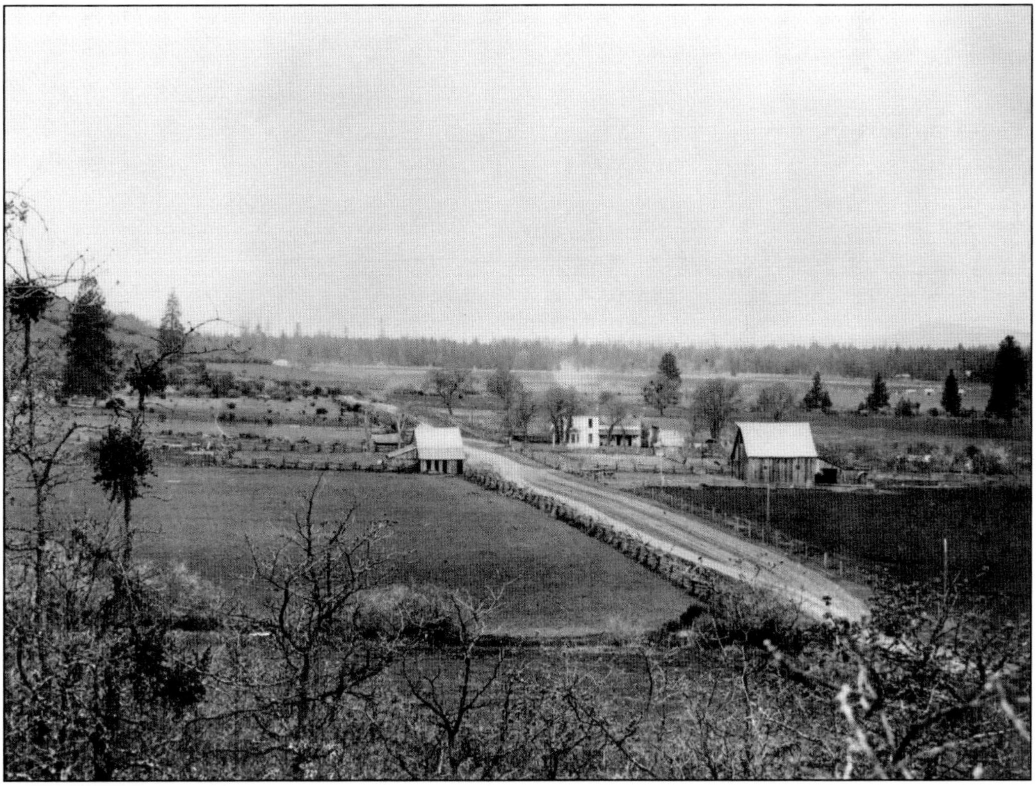

The Woodford and Mahulda Reames donation land claim was located northwest of Talent on what is now Colver Road. Reames was a skilled blacksmith with an allegiance toward the South during the Civil War. His hot temper and physical strength made him somewhat intimidating. The beauty and organization of his farm testified to his abilities as a farmer and blacksmith. (SOHS No. 8235.)

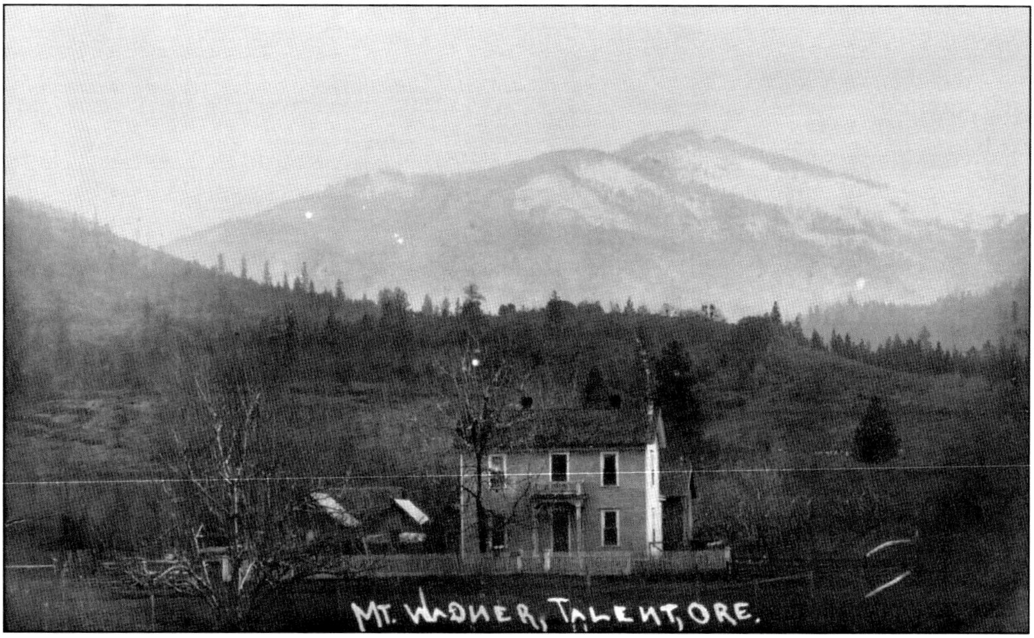

The Joseph Rapp family took over the James Thornton farm in Talent. Thornton moved to Ashland, where he and Jacob Wagner were partners in the flour and woolen mills. This home stood for many years along Rapp Road and became identified with the Rapp family even though it was probably built before they moved in. (Mike Cotta.)

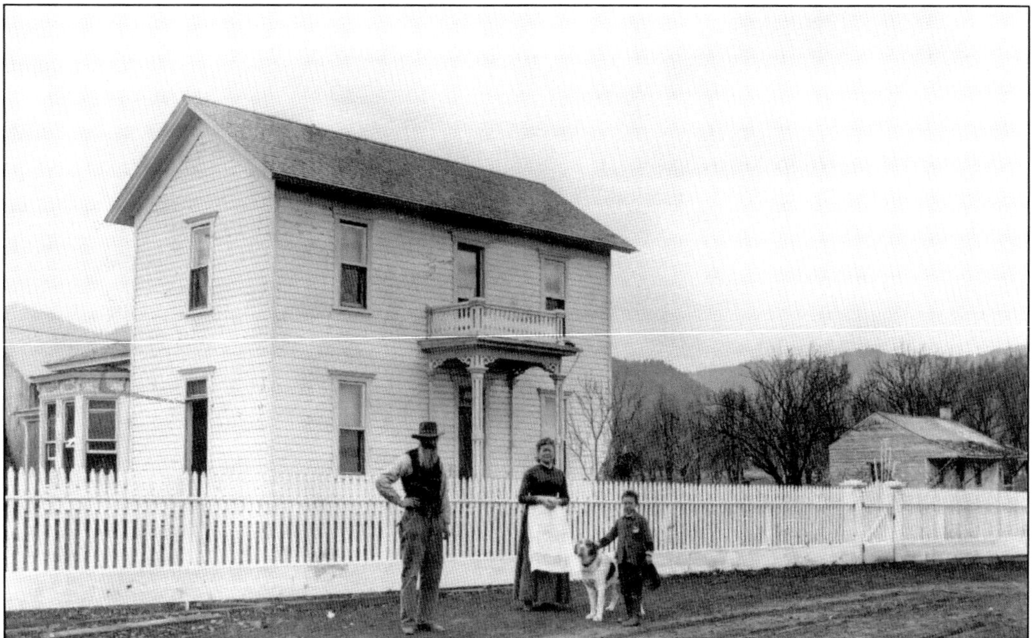

Martha Reames, a gregarious redhead, married a "sporty man," J. K. Ackley, before she settled down with her second husband, Joseph Rapp. Joseph was steady and conservative and was thought to be a member of the Church of the Brethren. Joseph, Martha, and son, Frederick, are seen in this photograph in front of their Rapp Road home. (SOHS No. 1931.)

This drawing was made from measurements and observations of the first-floor framing plan of the 1863 Welborn Beeson hexagonal house. Welborn's father, John Beeson, once boarded with and may have utilized designs from Orson S. Fowler, author of *The Octagon House: A Home for All, or a New Cheap, Convenient, and Superior Mode of Building.* (Larry Wikander.)

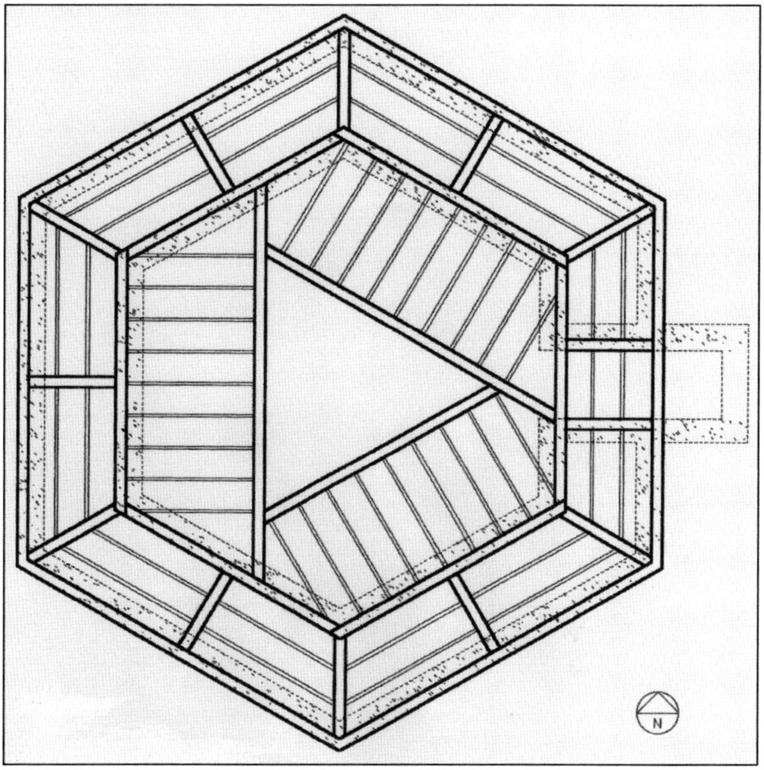

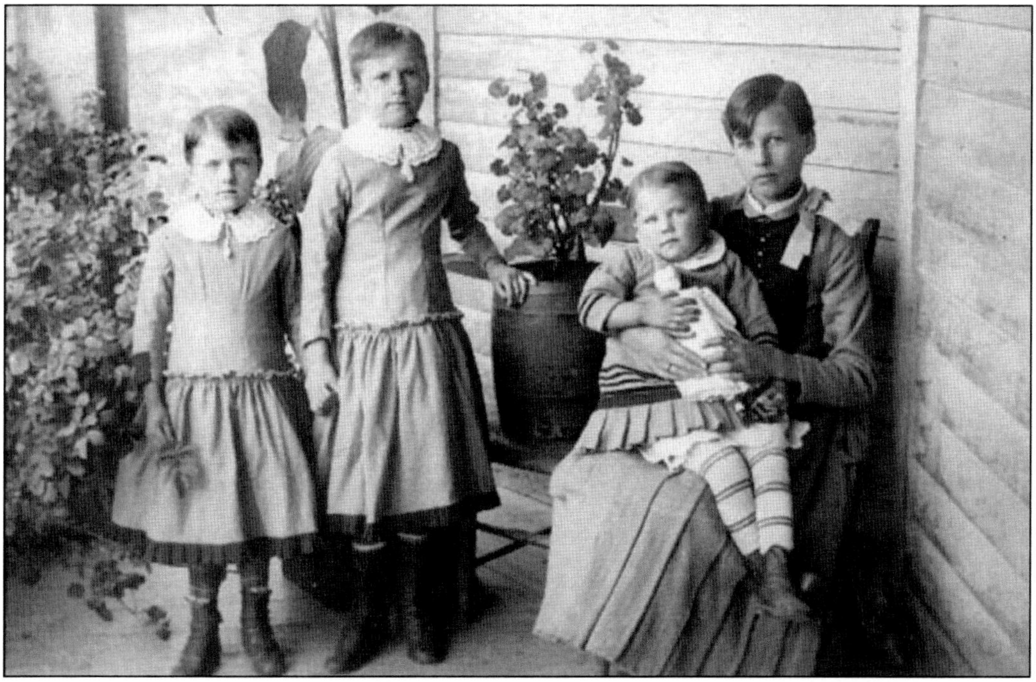

Anna, Frances, Winifred, and Jessie Beeson, daughters of Welborn and Kate Beeson, sat on the angled porch of their hexagonally shaped home on Wagner Creek Road in 1887. The unusually short hair on the girls might be attributed to their father's preference. (Carter family.)

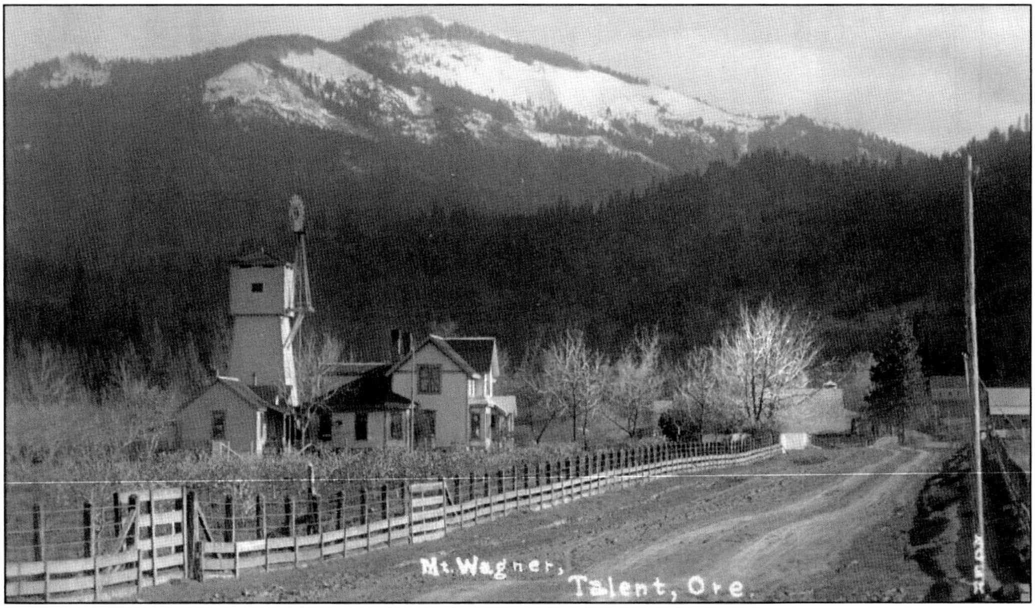

This house was once the only known hexagonal building in southern Oregon. It was built in 1863 when Welborn Beeson and his mother, Ann, lived on the property. Julius Foss bought it in 1890 and remodeled it, forever obscuring the original shape. A model of the hexagonal house is on display at the Talent Historical Society Museum.

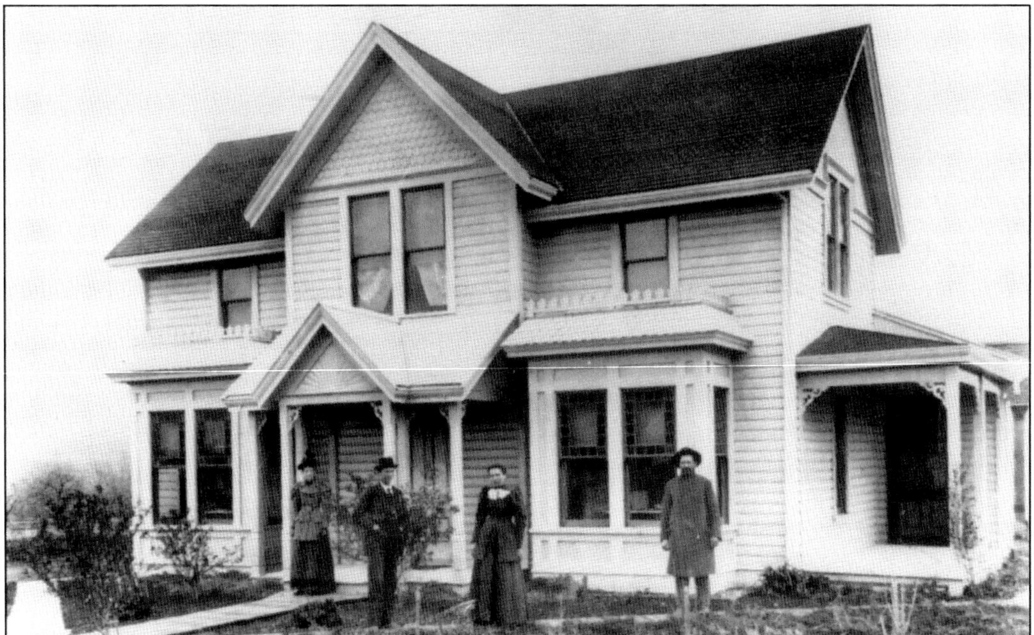

The Julius Foss family stood in front of the newly completed remodeled old Beeson house. Julius was able to buy the house and 25 acres with money from a winning Louisiana lottery ticket. He paid Welborn Beeson $4,000, which enabled Beeson to pay off all his debts and move his family across the street. Eventually one of the Beeson boys married Julius's daughter, Nellie. (SOHS No. 1092.)

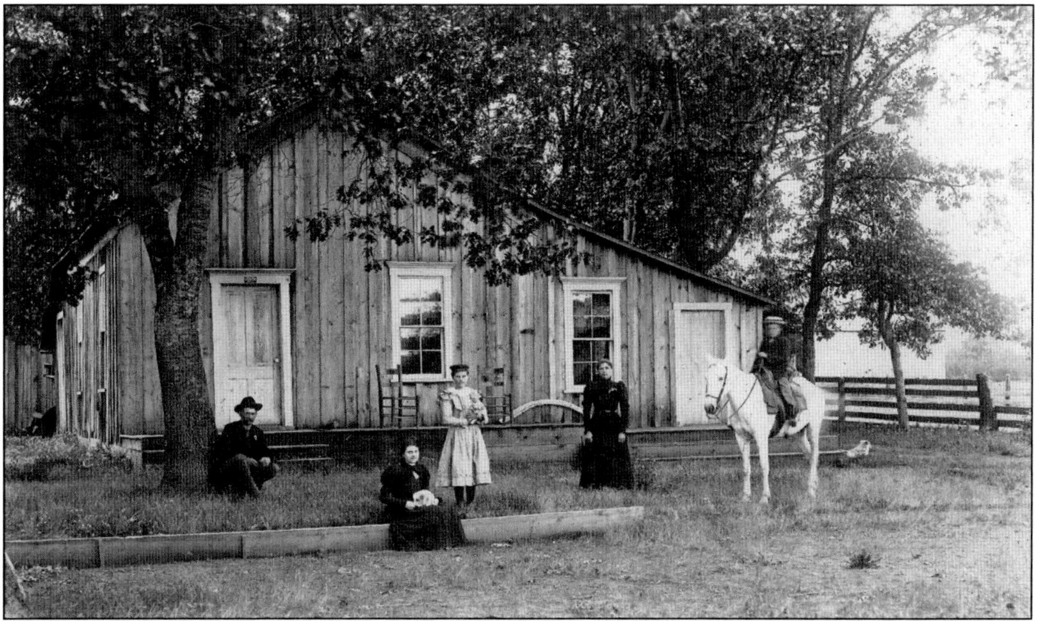

The house in this photograph behind Welborn Jr., Anna, Kate, Frances, and Carl Beeson was meant as a temporary structure put together from an old butcher shop. Welborn and Kate Beeson and their eight children moved to this house when debt and crop failure forced them to sell out. (Carter family.)

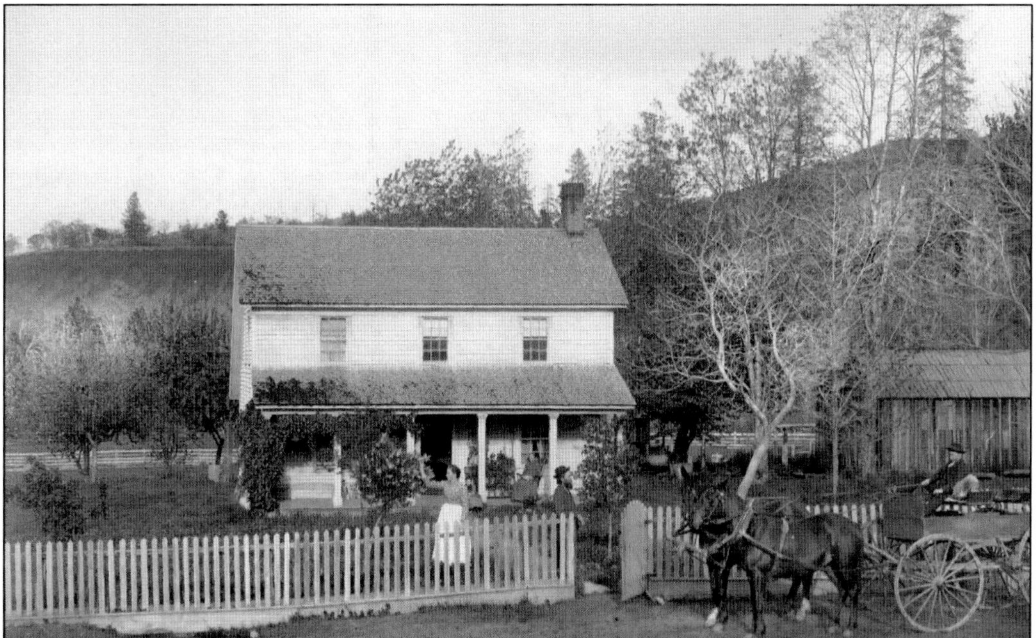

The John and Susan Robison family came from Iowa in 1853 and built this frame house in the late 1860s. This photograph was taken before 1889, the year Susan Robison (on the porch) died from injuries she received from being trampled by a horse. The swamp hickory tree on the right was planted by the Robisons and still stands to the side of the house today.

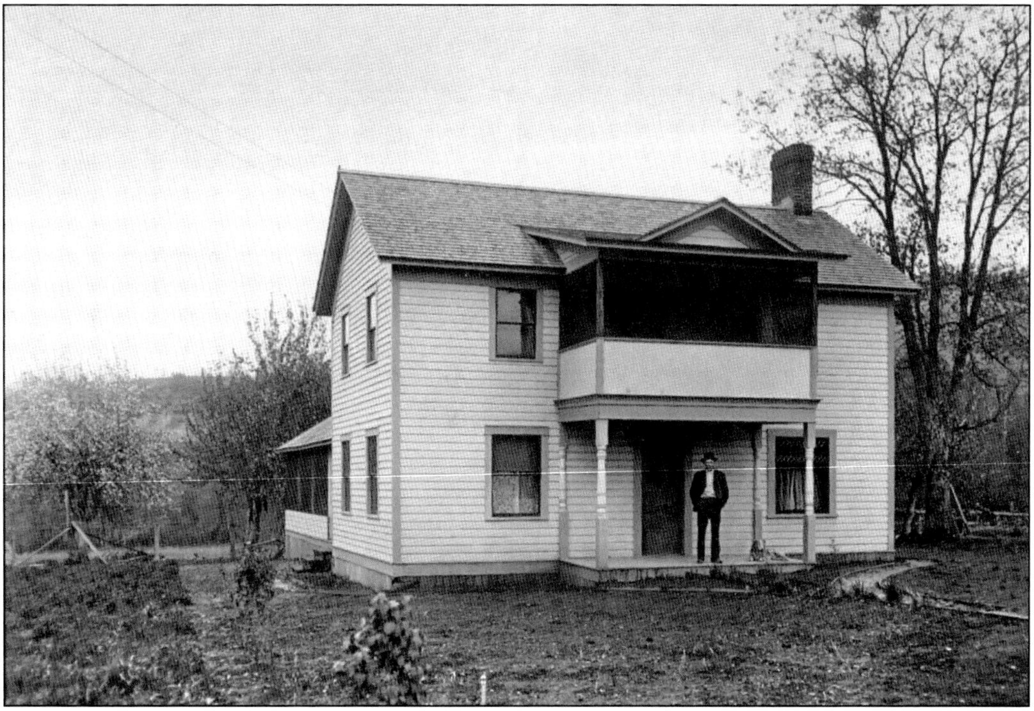

Edward "Eddie" Robison edited the *Talent News* from 1892 to 1894 with his stepfather, Willis J. Dean. He also used his press for dance cards, mining notices, and other printing jobs. The newspaper advanced the Populist cause and was very informative about local happenings. He is seen in this photograph in front of the remodeled home of his grandparents John and Susan Robison, which Edward inherited.

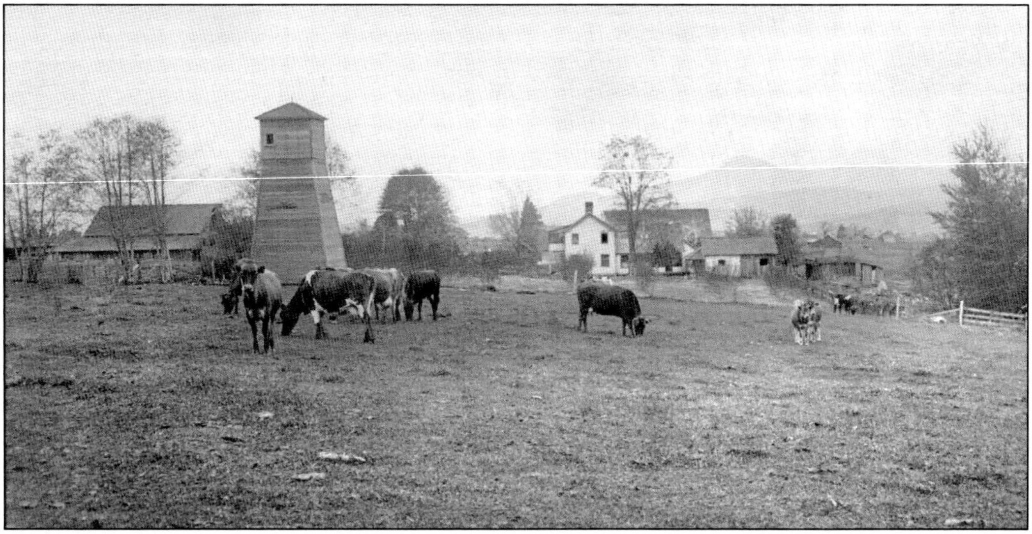

This photograph was taken looking northeast and gives a glimpse of the side of the Edward Robison dairy farm and house. Wagner Creek runs through the property on the right. The tower and outbuildings are gone, but the house is a private residence at 7103 Wagner Creek Road.

Anderson Creek Flume is still functioning on the old Samuel M. Robison homestead along the creek. Creek waters were used for irrigation, mining, and, at least once, for baptisms.

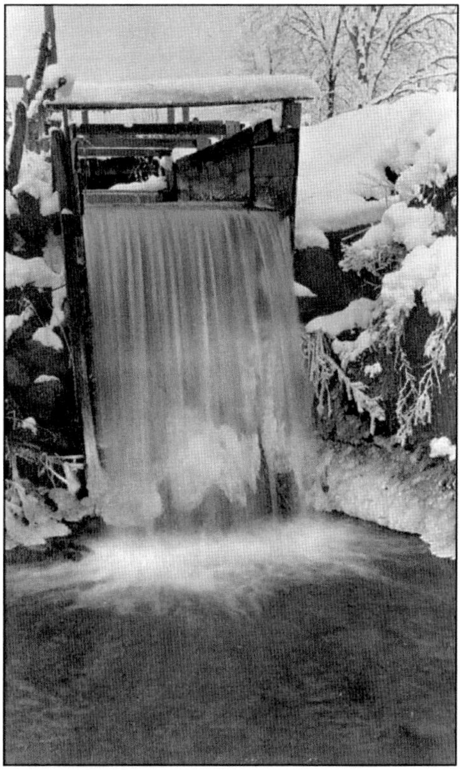

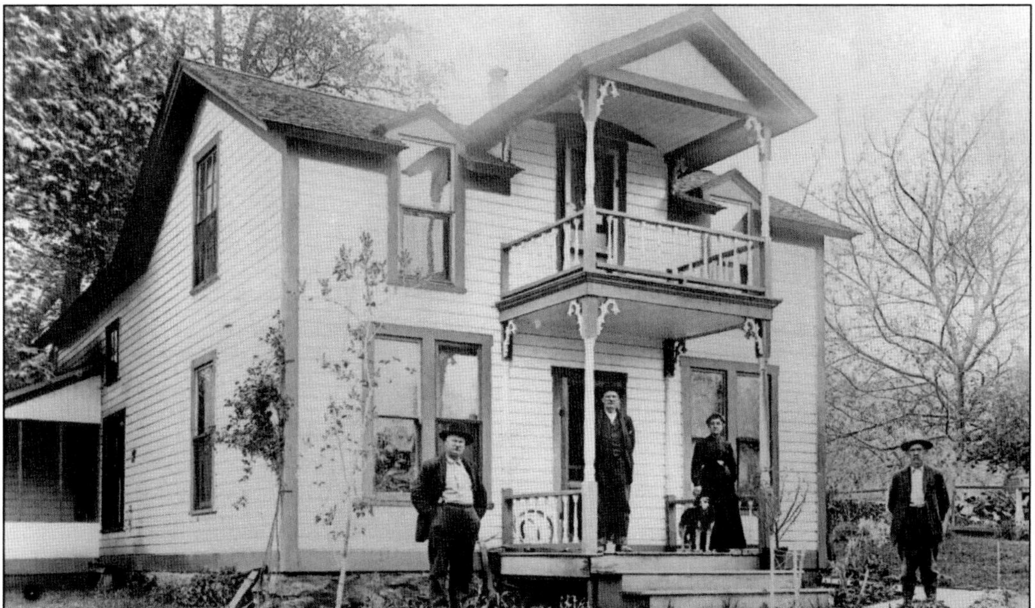

Samuel M. and Hannah (Barneburg) Robison were married in 1864. The neighborhood boys all performed a "chivaree" with noise and yelling to harass the new couple when they married. Samuel got so mad at the disturbance that he threatened to shoot them if they kept it up. This home was built for Samuel and Hannah Robison and still stands today off of Anderson Creek Road.

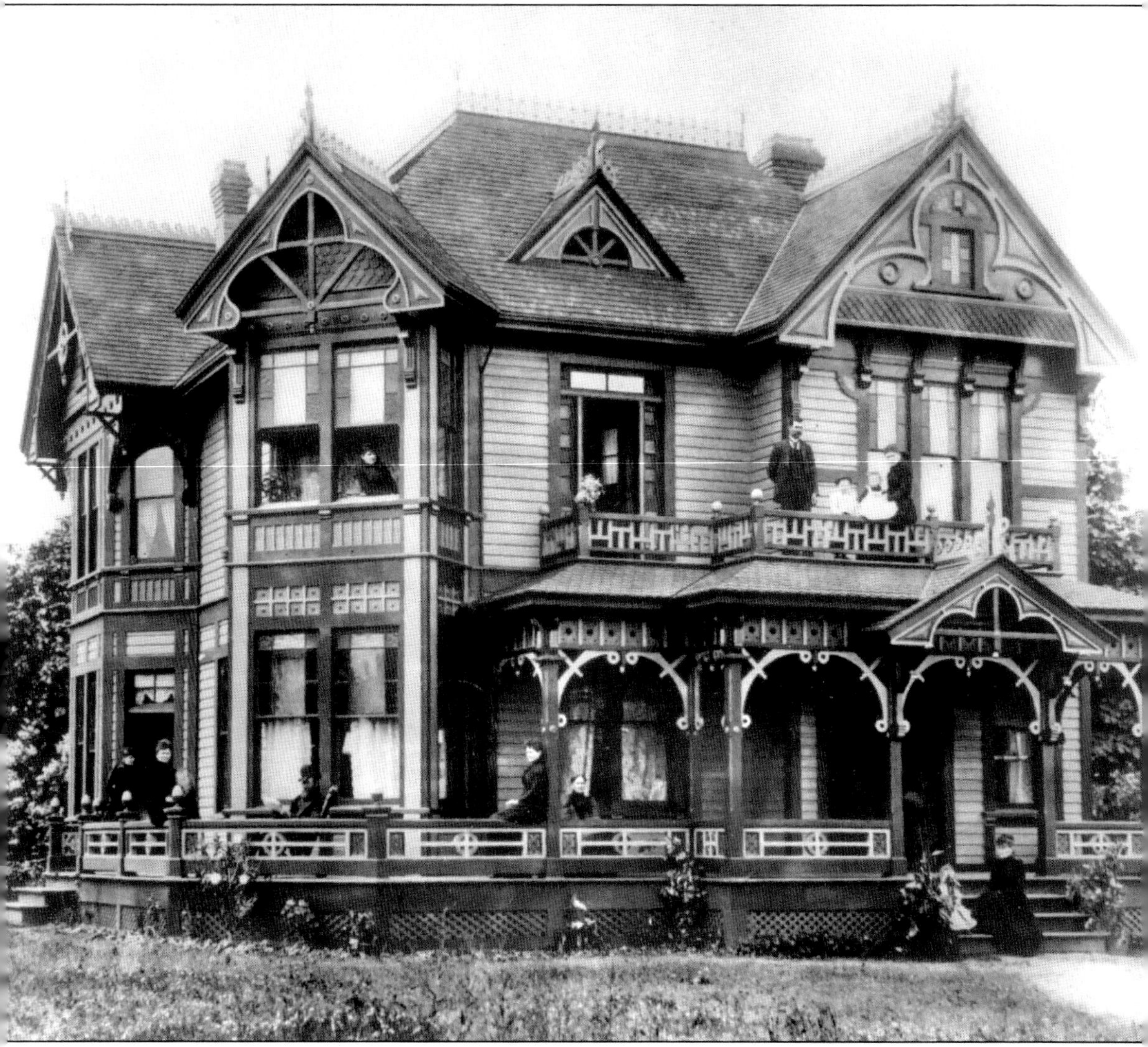

The elaborate gingerbread home of the Eli K. Anderson family was built along Colver Road. Anderson was very successful as a farmer, an orchardist, a miner, and a businessman in Ashland and Talent. He participated in the California Gold Rush and was lucky enough to have enough money to finance large mining operations in the mountains above Talent. The home burned down in the 1960s.

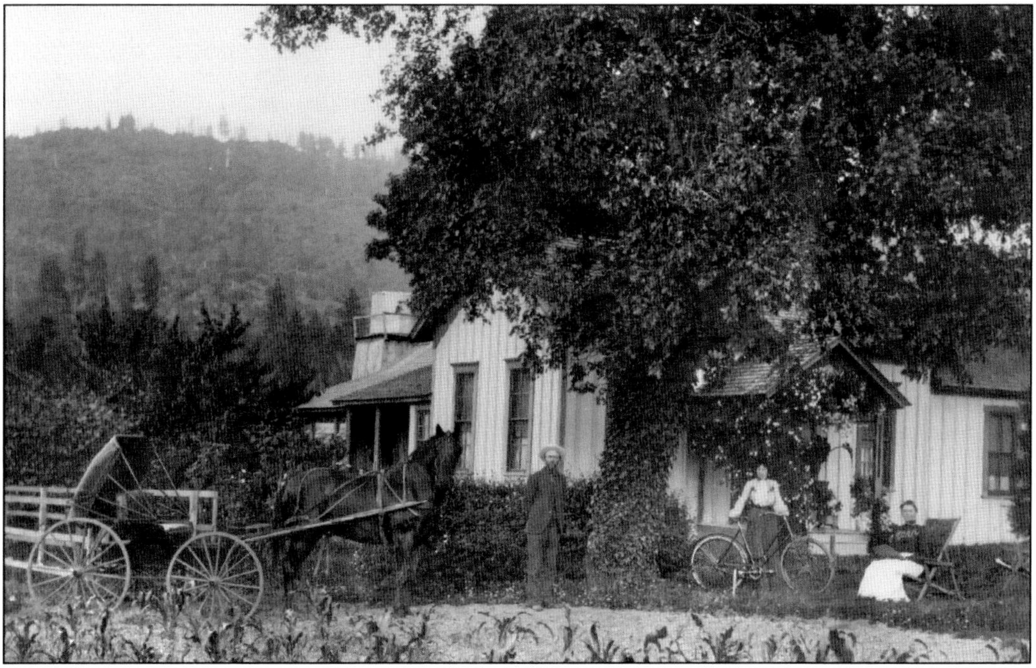

At first, Willis J. Dean was skeptical about spiritualism, but he encountered too many unexplainable occurrences to completely dismiss it. He experimented and finally concluded that contact with the dead was possible. He had a stand and a special table made to hold circle at this house on Wagner Creek Road. Pictured in front of the house are Dean; his wife, Ursula; and stepdaughter, Minnie Robison. (Ben Truwe.)

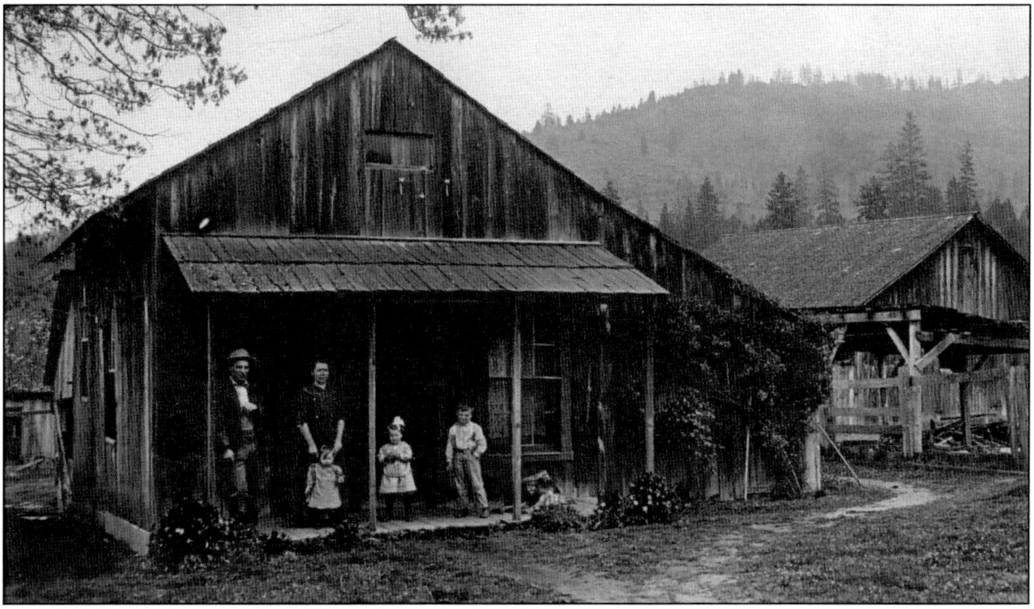

In 1893, the home of James and Arminda Stearns Purves burned down. Neighbors helped them construct a temporary home out of a wooden storage building on the same property. This photograph taken in 1912 is of the post-fire home with James's son, Robert Purves, and his family on the porch. (Maxine Kincaid.)

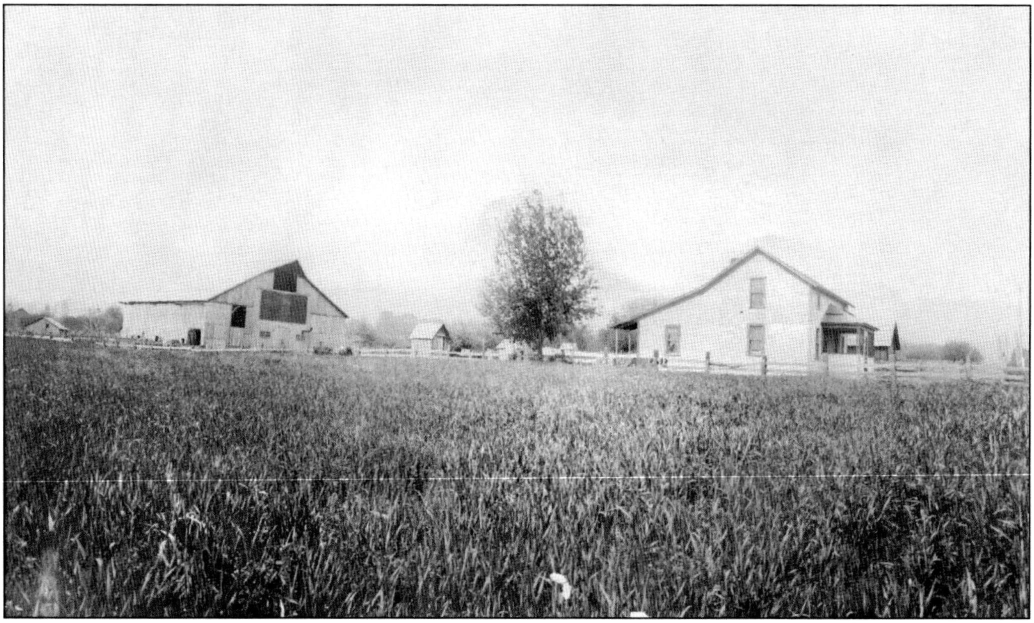

Three Holdridge families came to Wagner Creek in 1908 from Iowa and bought acreage along Rapp Road. They came just in time to take advantage of the orchard boom. The corner of Rapp and Wagner Creek Roads became known locally as Holdridge Corner.

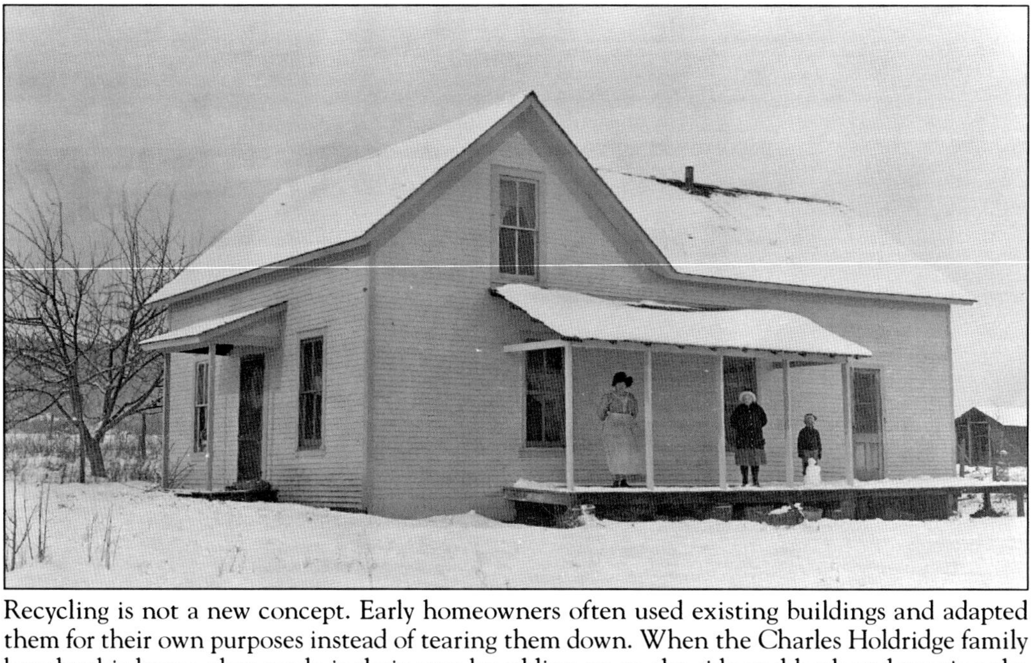

Recycling is not a new concept. Early homeowners often used existing buildings and adapted them for their own purposes instead of tearing them down. When the Charles Holdridge family bought this home, they made it their own by adding on to the side and back and turning the house around to face the road.

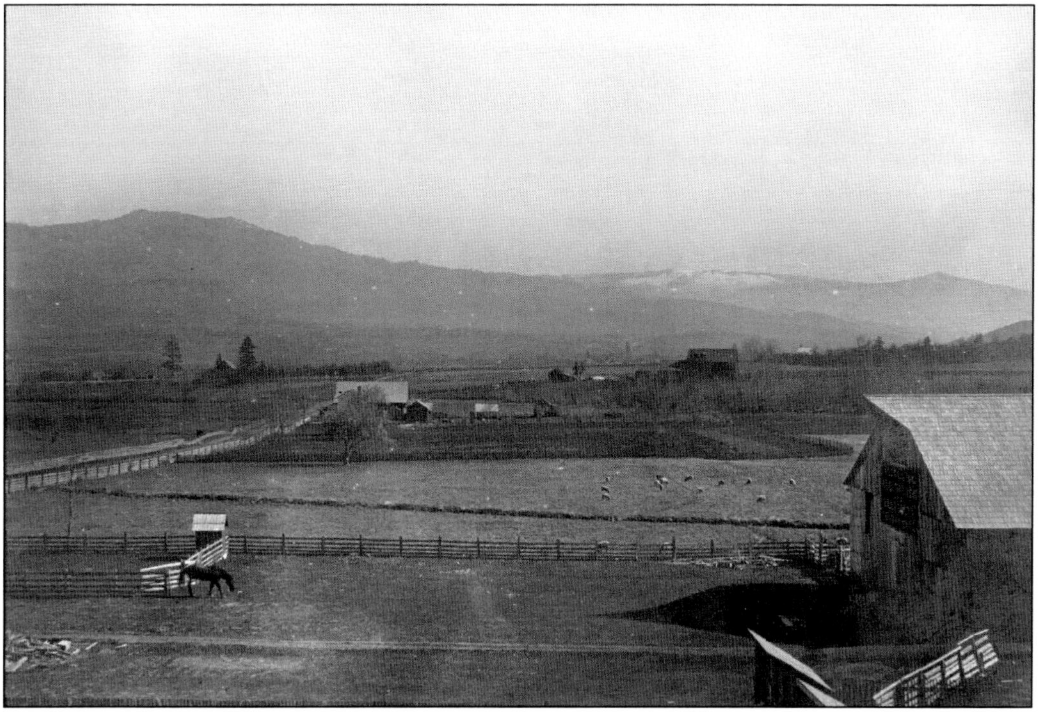

Joseph and Martha Rapp and son Frederick established this farm along Rapp Road and began the tradition of growing the most delicious watermelons around. Nearly every boy in Talent at some time attempted to steal from the patch and dodge Rapp's buckshot. Frederick Rapp's uncanny sense of when he was being robbed was as legendary as his watermelons.

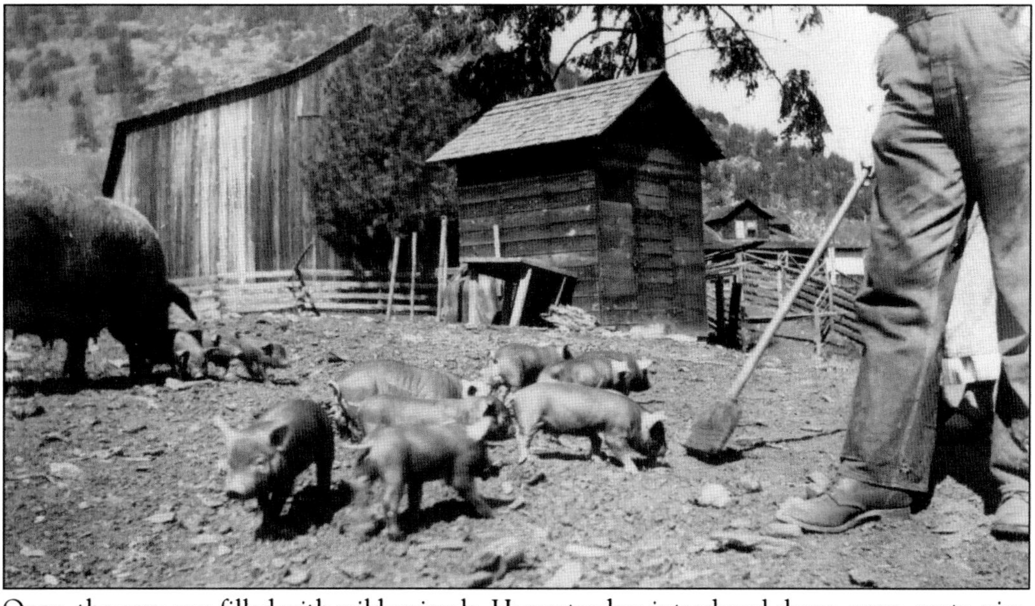

Once, the area was filled with wild animals. Homesteaders introduced sheep, cows, goats, pigs, chickens, horses, and pets that competed with the wild animal habitat. George Kerby made a special herding devise out of a broom to push the piglets around on his Yank Gulch property. (Kerby family.)

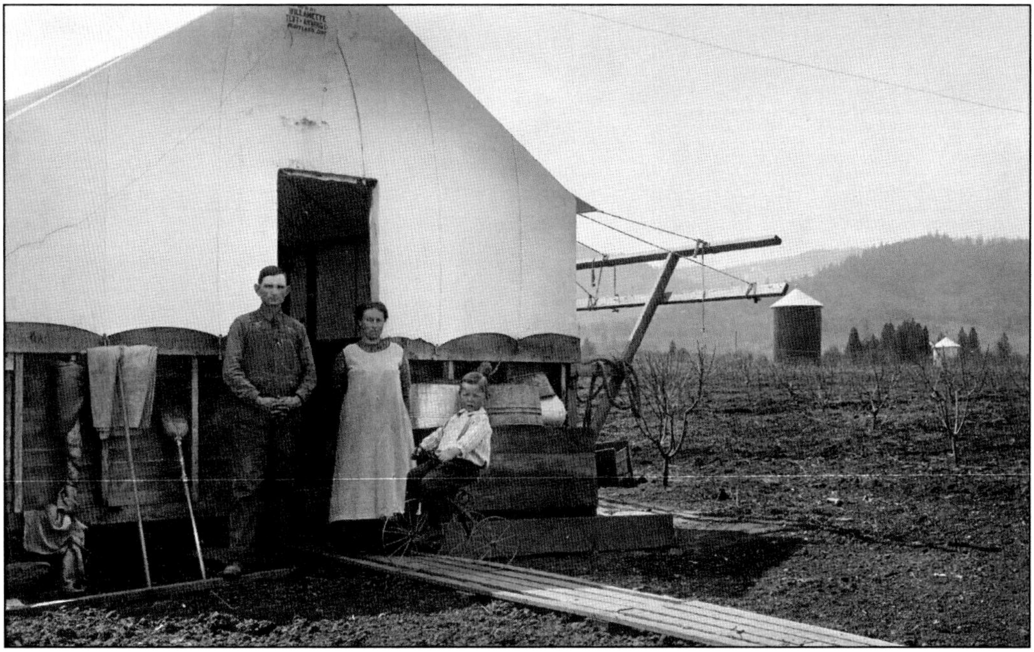

William and Ethel Childers and their son, Buford, spent the summer of 1911 in this tent. Though their circumstances were humble, Buford could still ride his tricycle and his father, William, wore a tie with his overalls.

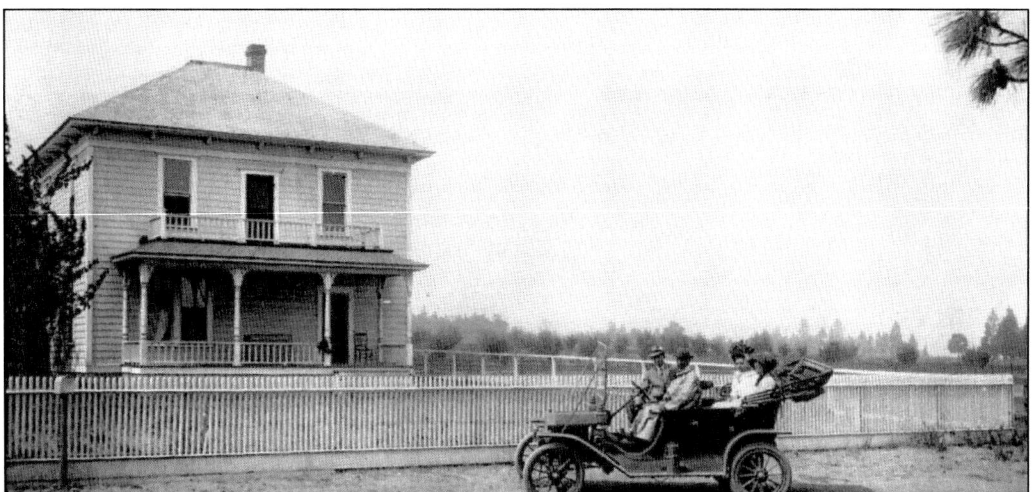

Built about 1906 by Emmett Beeson, this house became known as the Gleim family home. Harry S. Gleim and his wife came from El Paso, Texas, to Talent to get in on the pear rush. They had heard that the pears would bring as much wealth as the gold had in pioneer times. They loaded up their car and moved, though it took them several years to completely sever ties with Texas. (Harry Gleim.)

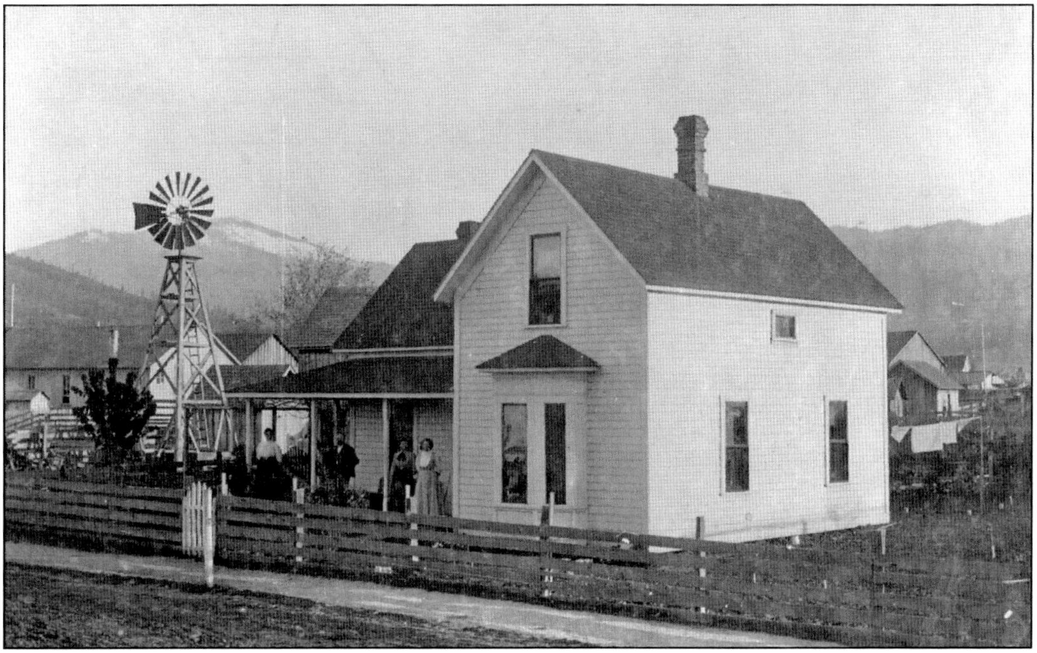

The Bell house was built about 1897 and is one of the oldest houses still standing in Talent. Thomas Jefferson Bell and his second wife, Amanda, bought the house in 1906 and converted it to a boardinghouse. Amanda did the cooking, Thomas tended the garden, and a hired girl did most of the cleaning. A grapevine over 100 years old still graces the side porch.

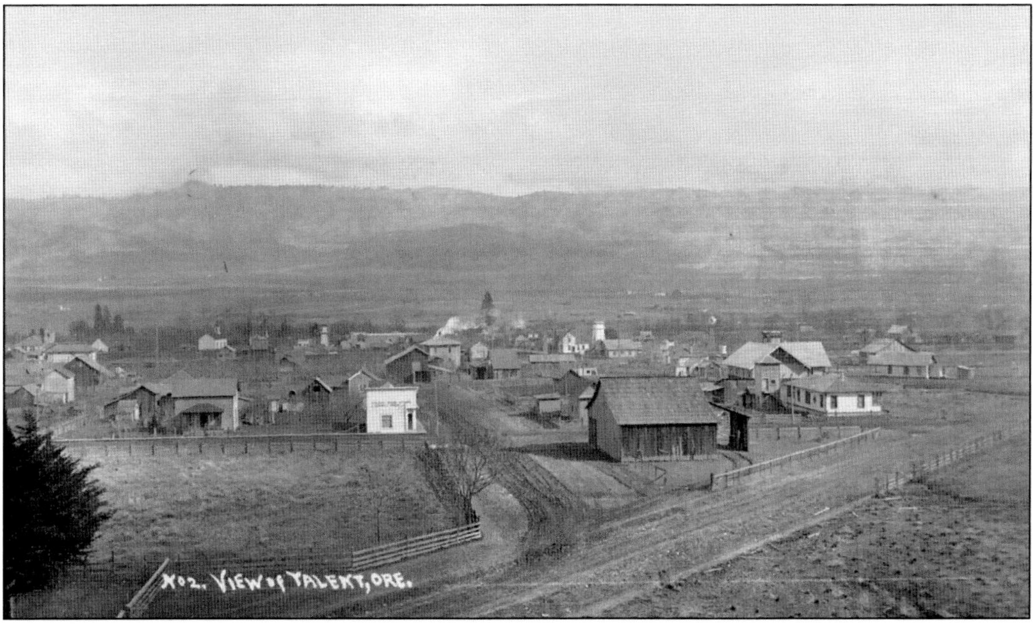

The city of Talent has always been on the small side. The citizens consistently say they like it that way. Ordinances have rid the city limits of barns and outhouses and paved the streets, but the basic borders remain the same. This photograph was taken about 1912 from the brick schoolhouse and shows West Wagner Street in the foreground intersected by Bain Street in the center. (Mike Cotta.)

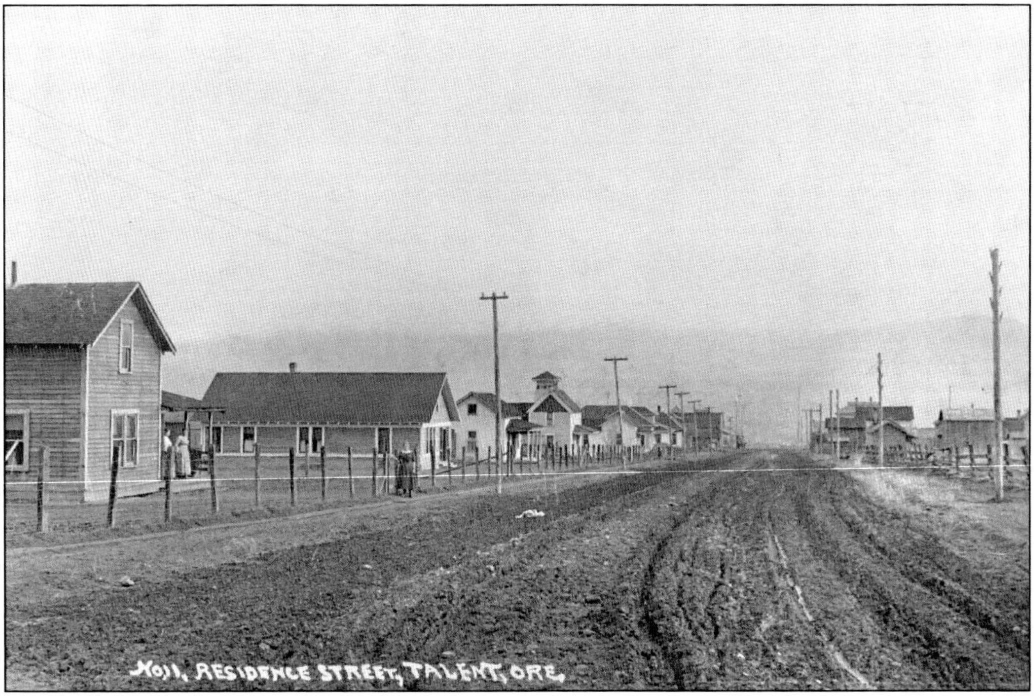

These Main Street houses were all fairly new when this photograph was taken, and some of them still stand today. Electricity and phone lines had made their way to Talent, but the roads were still unpaved.

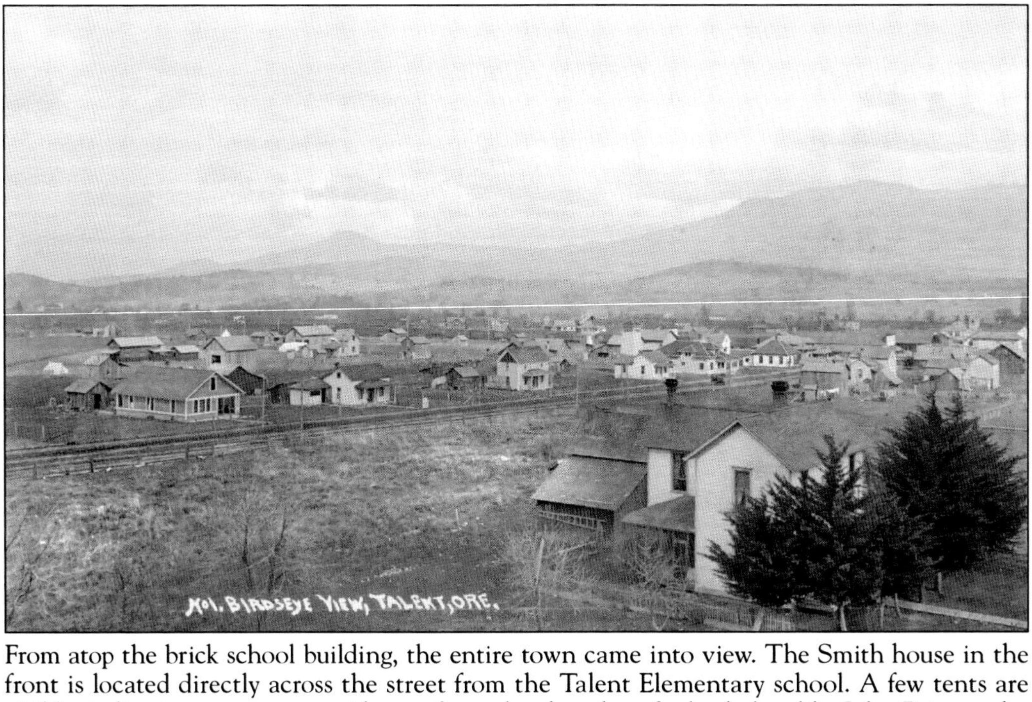

From atop the brick school building, the entire town came into view. The Smith house in the front is located directly across the street from the Talent Elementary school. A few tents are visible, indicating temporary residences for orchard work or for bachelors like John Briner, who liked the freedom of moving his tent from place to place. (Mike Cotta.)

Four

Downtown

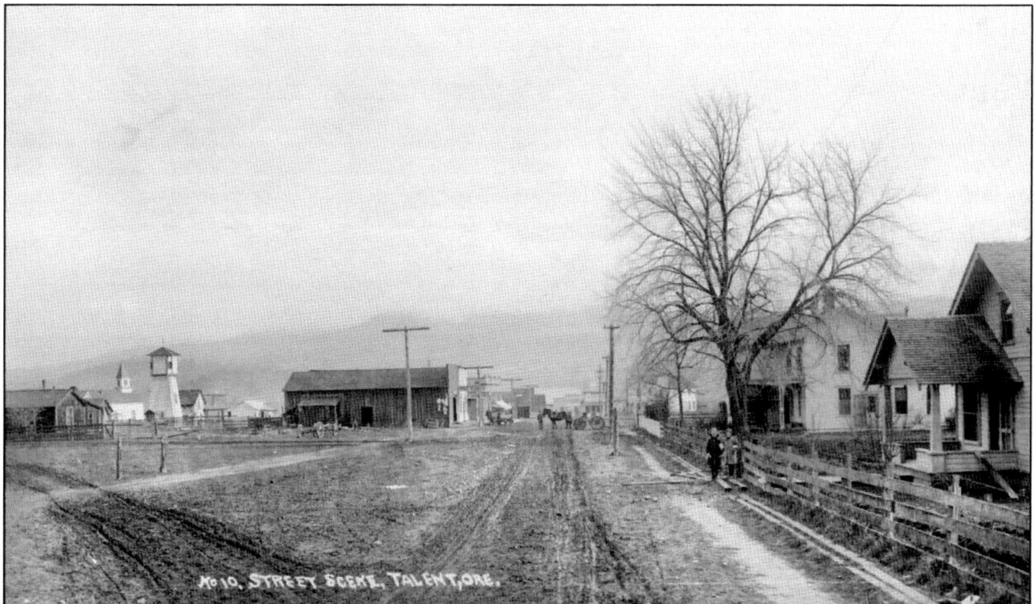

Like many American towns, Talent made a slow transition from dirt to paved roads, from buggies to automobiles. This photograph was taken about 1912 from the railroad tracks on Wagner Street looking toward Talent Avenue. The wooden building in the center was a blacksmith shop located on the site of today's Tarks Market. (Mike Cotta.)

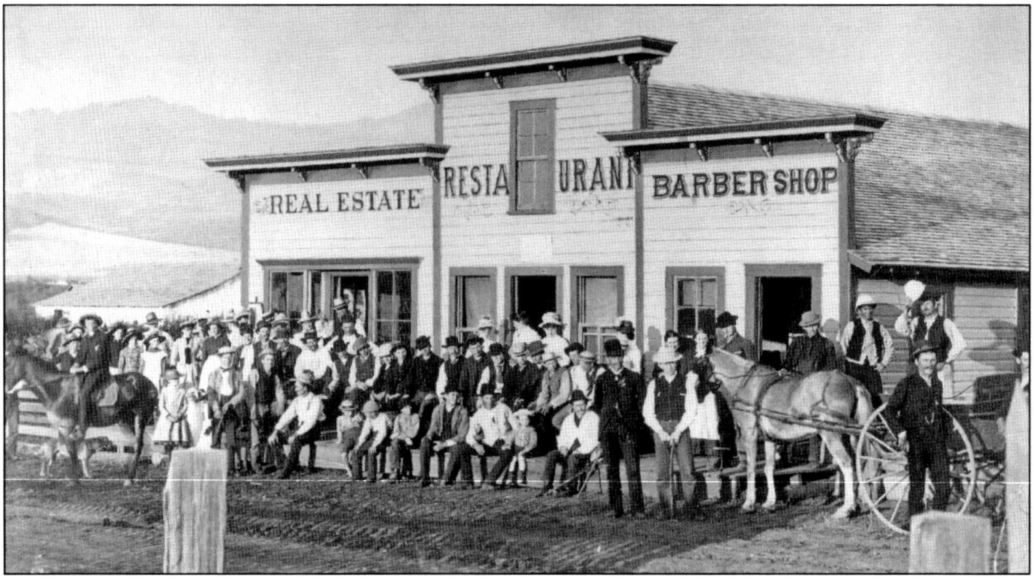

Nearly half the town showed up for what might have been the opening day of the Sherman block in 1888. Salisbury Sherman, a Civil War veteran, built a real estate office, restaurant, and barbershop with three bedrooms above the restaurant for boarders. The buildings reflected the town's new prosperity with the coming of the railroad. (SOHS No. 1076.)

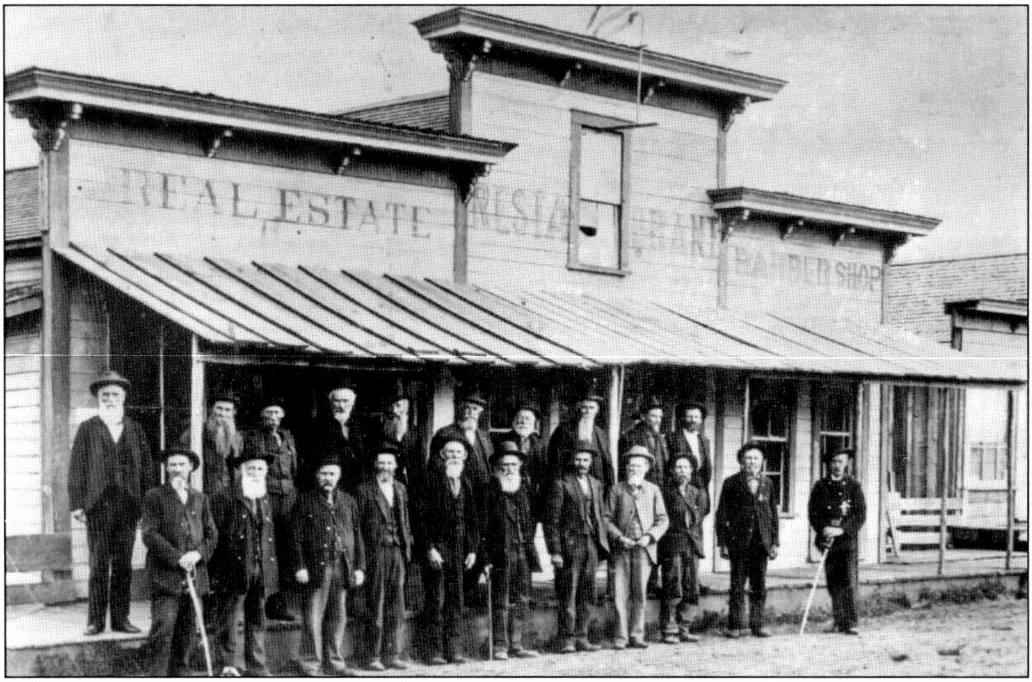

Civil War veterans lined up in front of Talent's first business district, built in the late 1880s. Their stories and experiences fired up the youngsters and old-timers alike. They belonged to the Grand Army of the Republic, which had regular meetings in Ashland, Talent, and Phoenix, and were always on hand to march in the Fourth of July parades.

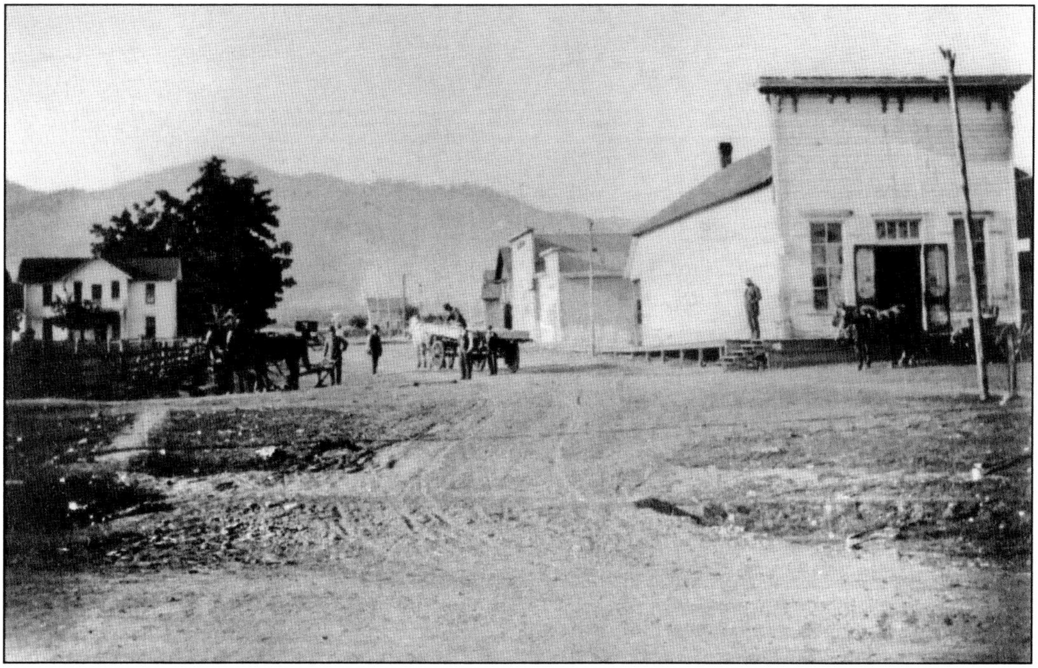

When this picture was taken in 1907, Charles W. Wolters owned many of the businesses in Talent. His house, seen on the left, was conveniently located for keeping an eye on the stores. The house still stands today, but none of the commercial buildings do. Charles died in 1912, and his widow, Amanda Olive Wolters, lived in the house until she died in 1921.

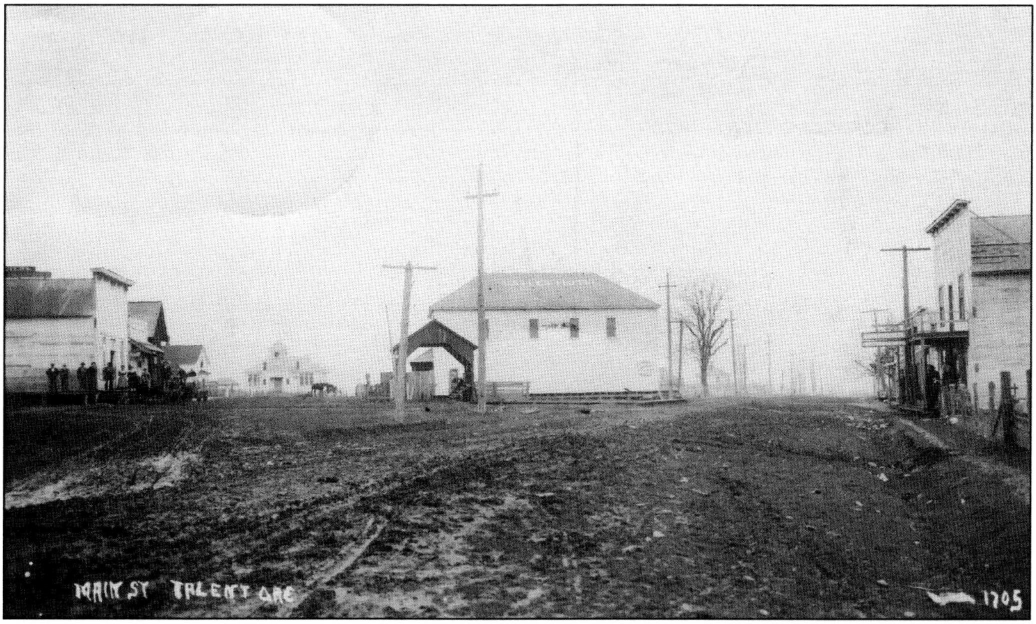

Along the old stage route on the right (now Talent Avenue) was Joseph Spitzer's livery stable and a grocery store operated by Spitzer's son, William. A platform on the wooden sidewalk between the livery stable and the grocery store was a gathering place for a group of old men called the "Spit and Argue" Club. The large building in the center was a hotel and dance hall.

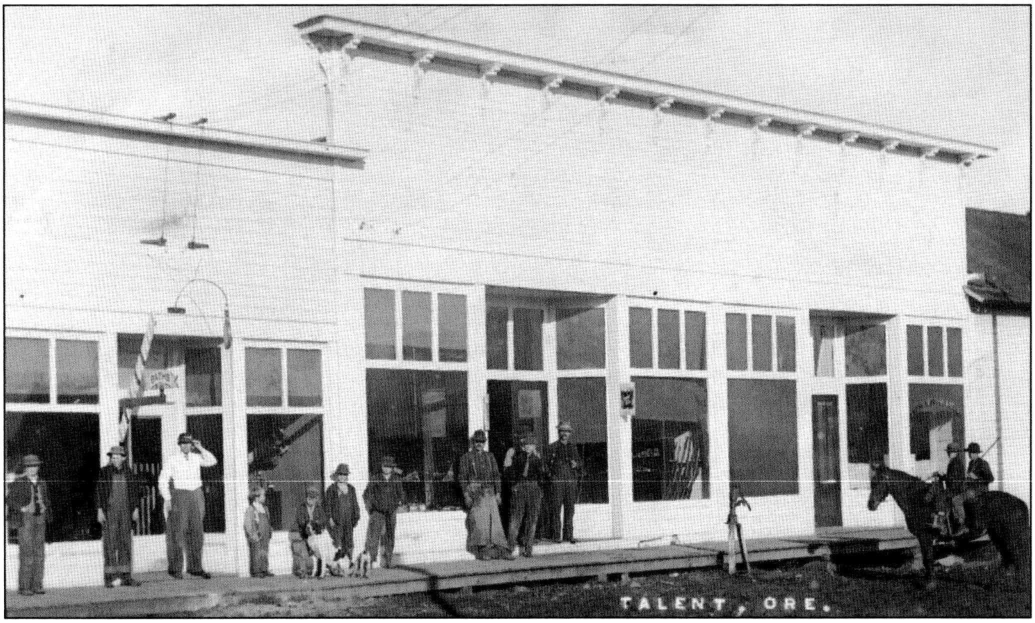

The storefronts on Wagner Street were curiously unmarked around 1910. Perhaps it was completely unnecessary to name the buildings that everyone in town knew so well. The building on the left was most likely J. F. Norman's Barbershop with the striped pole and the "Baths" sign over the door. Baths were commonly offered in early barbershops. (Mike Cotta.)

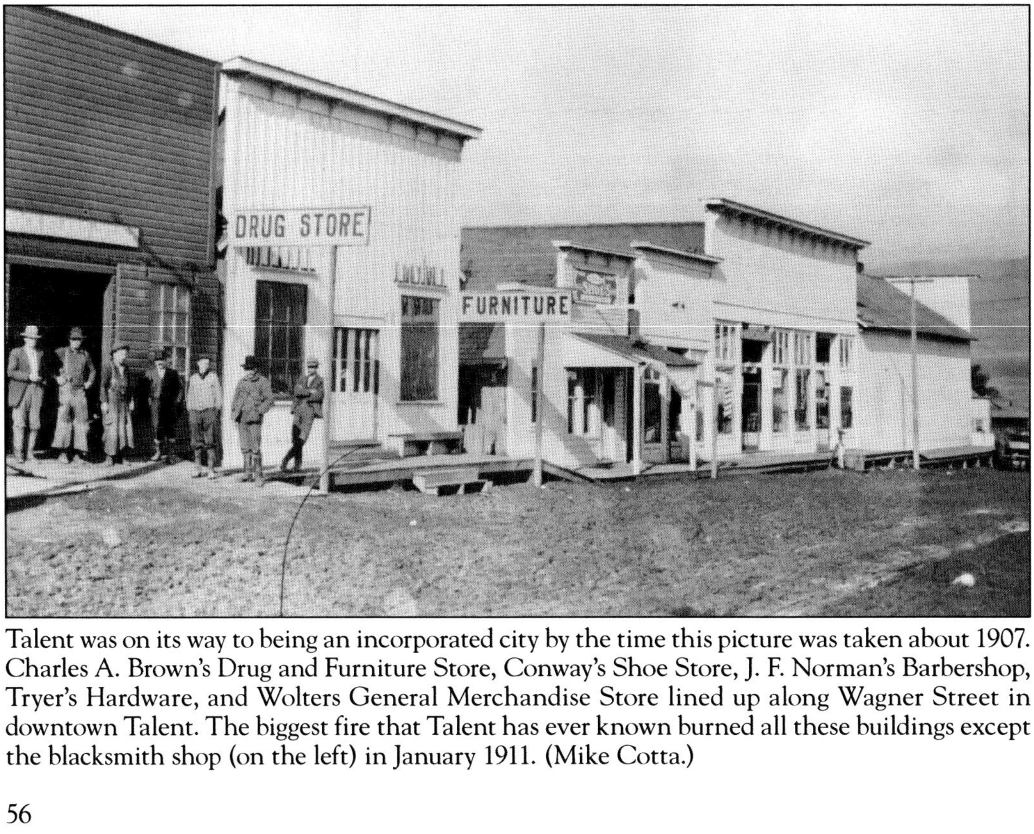

Talent was on its way to being an incorporated city by the time this picture was taken about 1907. Charles A. Brown's Drug and Furniture Store, Conway's Shoe Store, J. F. Norman's Barbershop, Tryer's Hardware, and Wolters General Merchandise Store lined up along Wagner Street in downtown Talent. The biggest fire that Talent has ever known burned all these buildings except the blacksmith shop (on the left) in January 1911. (Mike Cotta.)

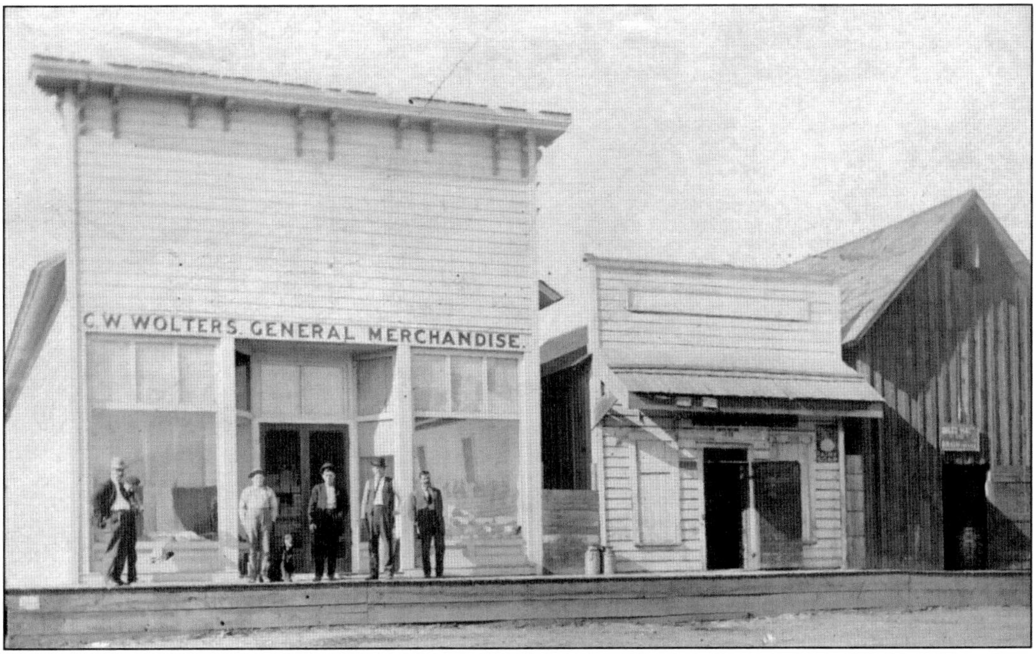

Charles W. Wolters changed the windows in the front of this store to entice window shoppers. The first version of the store had a flat front and smaller eight-paned windows. Chester Wolters is seen in this photograph standing beside the dog, and his father, Charles W. Wolters, store owner, is second from the right. (Jerald McGrew.)

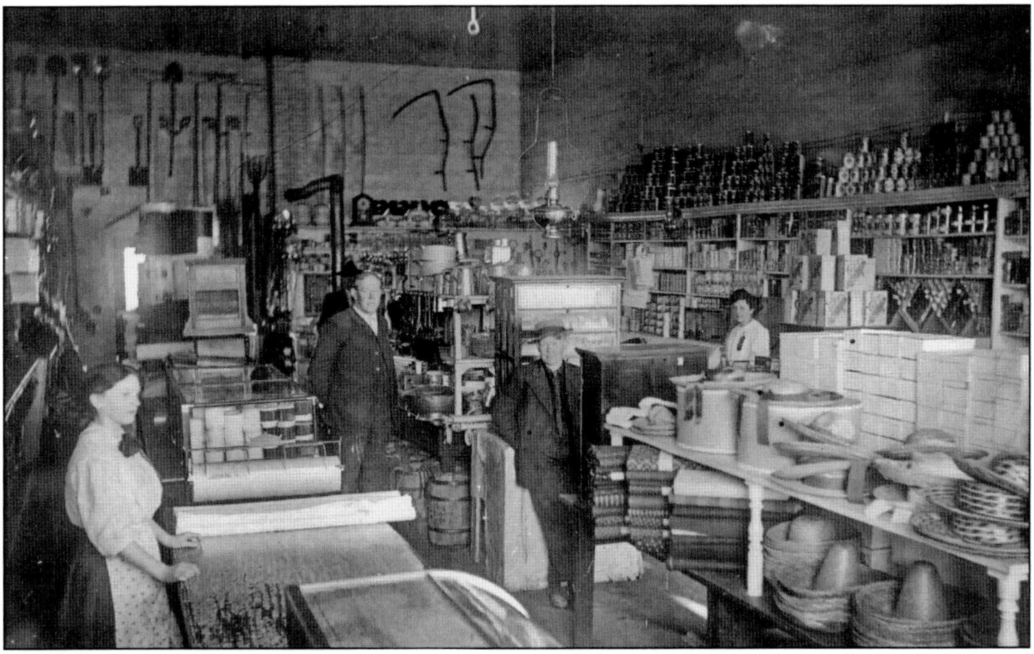

In 1906, Katie Lamb started working for $3 a week at the Charles W. Wolters General Merchandise Store, and it was part of her job to arrive at 6:00 a.m. to build a fire in the woodstove. This image shows the original Wolters store interior, where Katie worked before the 1911 fire. (SOHS No. 1084.)

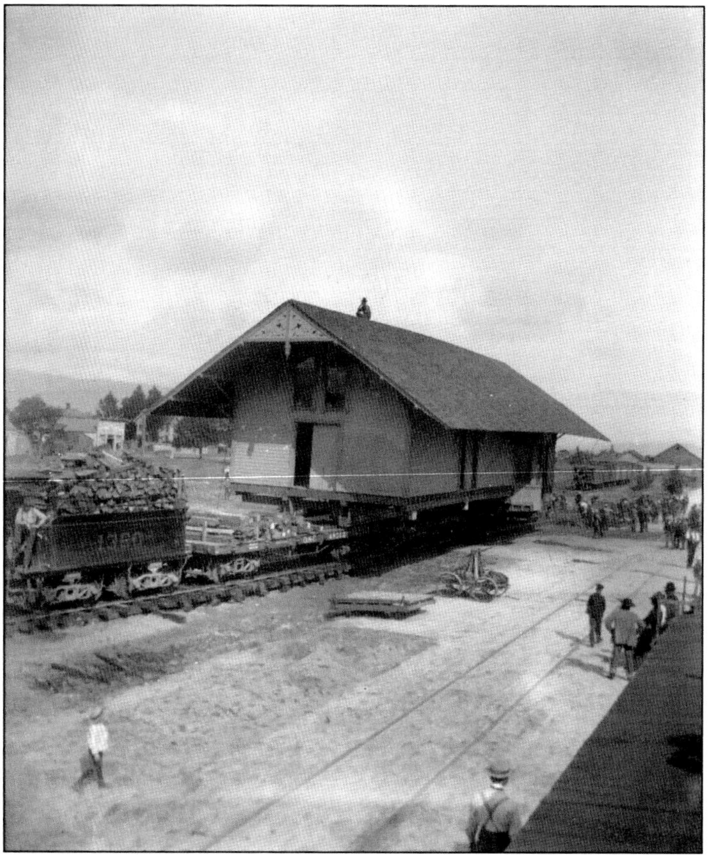

Talent residents spent years lobbying for a depot and side track after the railroad came through in 1884. Finally in 1900, the used Southern Pacific Station from Medford came to Talent by rail already assembled. This image shows the depot on a flatcar at Medford ready to be transported to its new home in Talent. The depot building that stands today is a reproduction of the old one. (James Uerlings.)

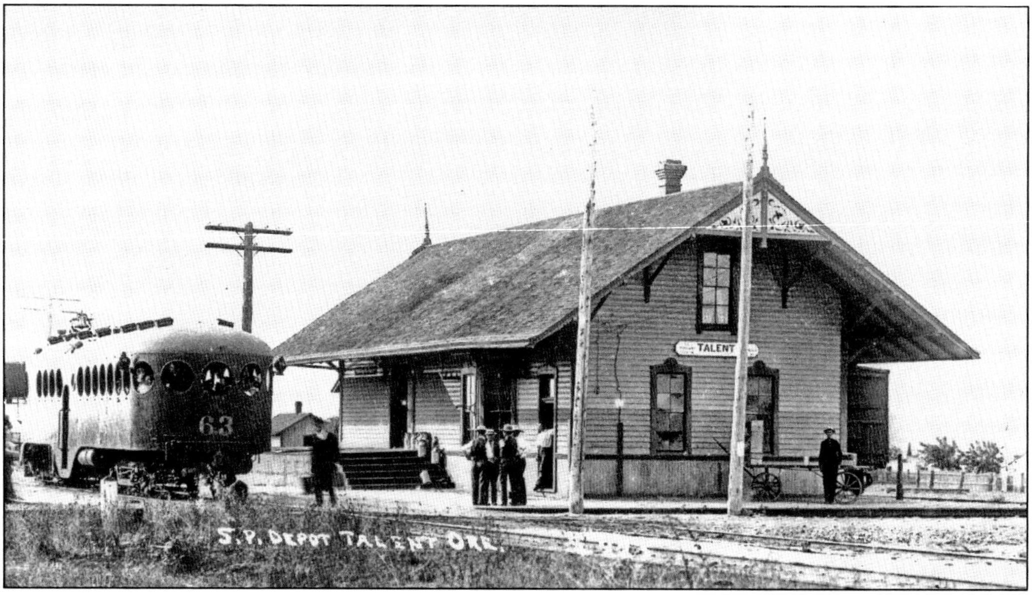

The lightweight McKeen Motor passenger car with round windows was gasoline powered, quieter, and less costly than the regular passenger trains. It ran from Grants Pass to Ashland and was manufactured between 1907 and 1913. (Terry Skibby.)

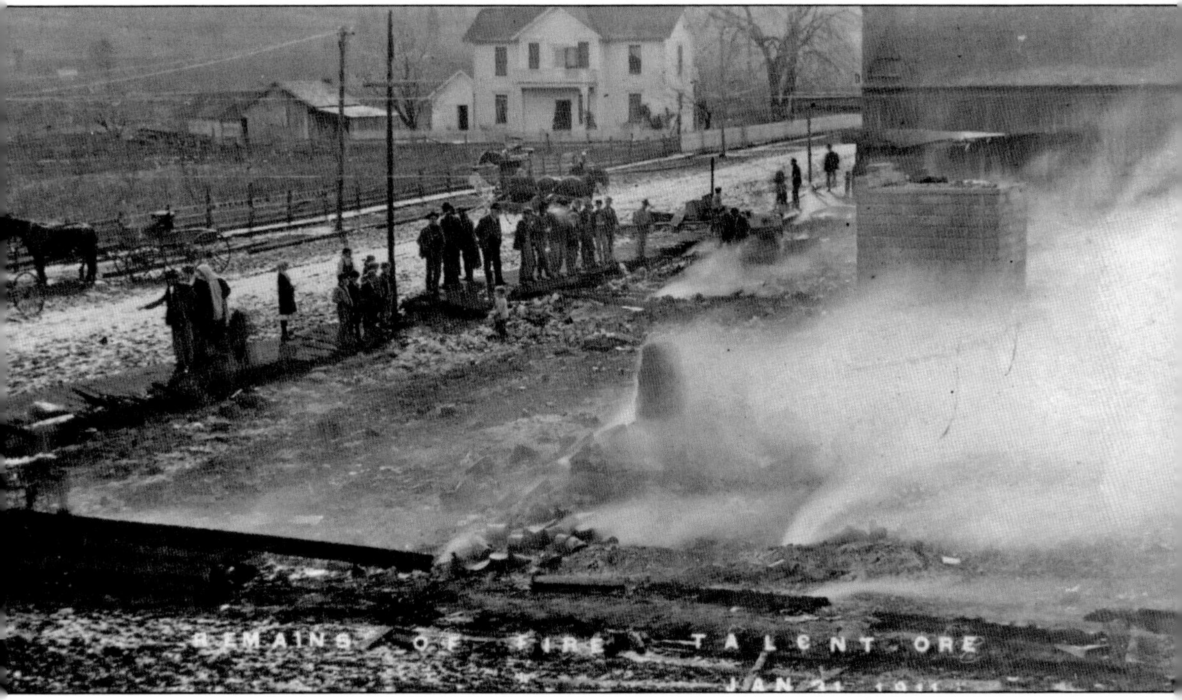

In January 1911, a fire destroyed the main business district on Wagner and I Streets in Talent. A loss totaling about $30,000 took a shoe shop, hardware store, barbershop, general store, and bank—only the vault from the bank survived and the contents were spared. The blacksmith shop seen in the upper right of the photograph was not affected.

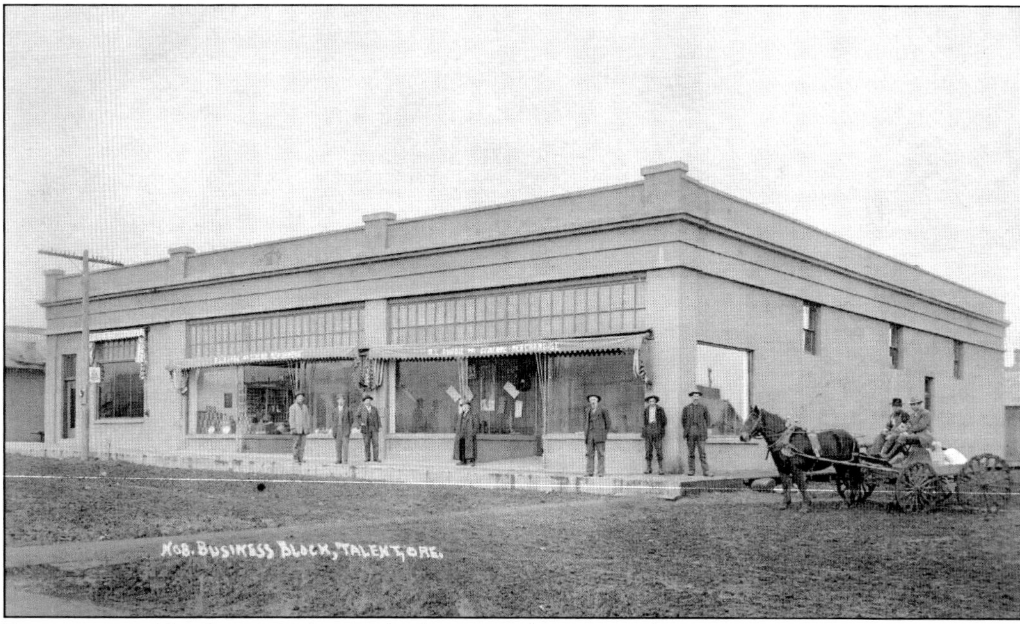

Businessman Charles W. Wolters, who lost up to $14,000 in the fire, wasted no time in replacing the burned structures with a modern concrete building. This photograph was taken shortly after the new store and bank were built. To bring customers into the bank, a full-page advertisement offered six reasons why people should use banks instead of hiding money under the mattress. (Jerald McGrew.)

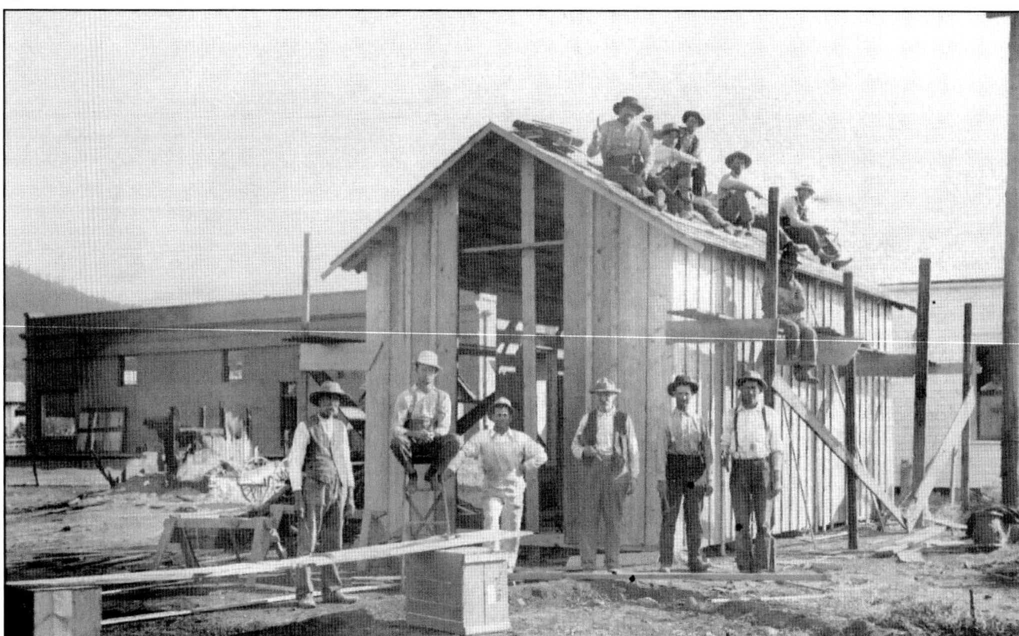

No shortage of labor existed in Talent after the 1911 fire. Ashes and charred remains of the wooden buildings can be seen in the background along with the new concrete Wolters store already in place. The structure being constructed was probably a storage building along what was then I Street (now Market Street) across from the Bell house. (Everett Taylor.)

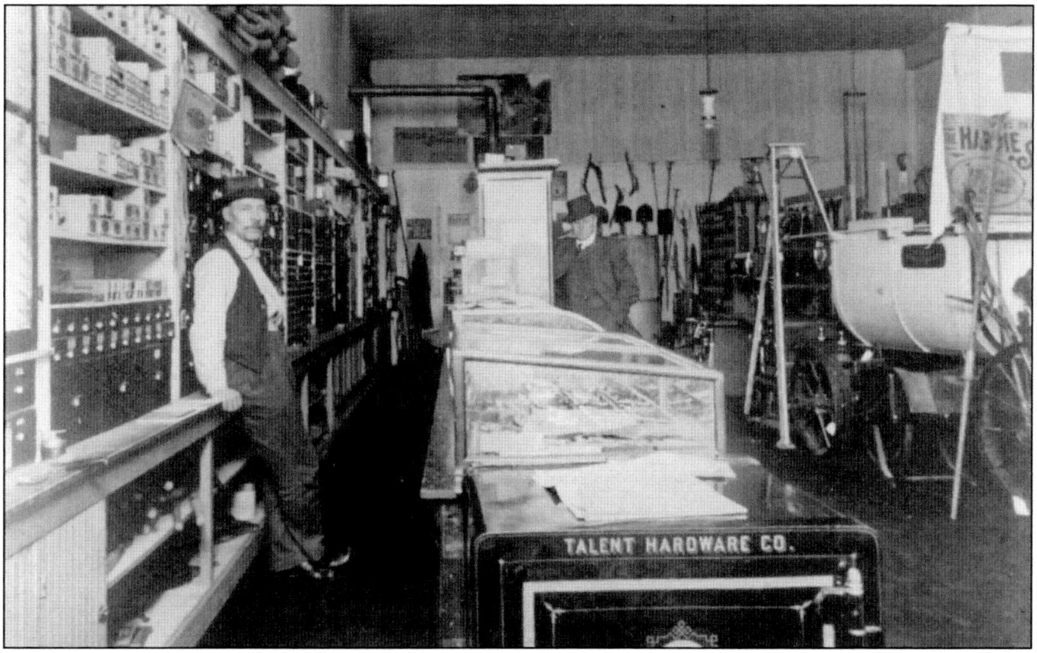

The Talent Hardware Company was owned by Marion Tryer, Talent mayor from 1911 to 1912. He offered spraying machines, paint, tinware, silverware, tools, and stoves for sale. The current Talent Ace Hardware store on Highway 99 offers plumbing, electrical, and garden supplies and liquor sales.

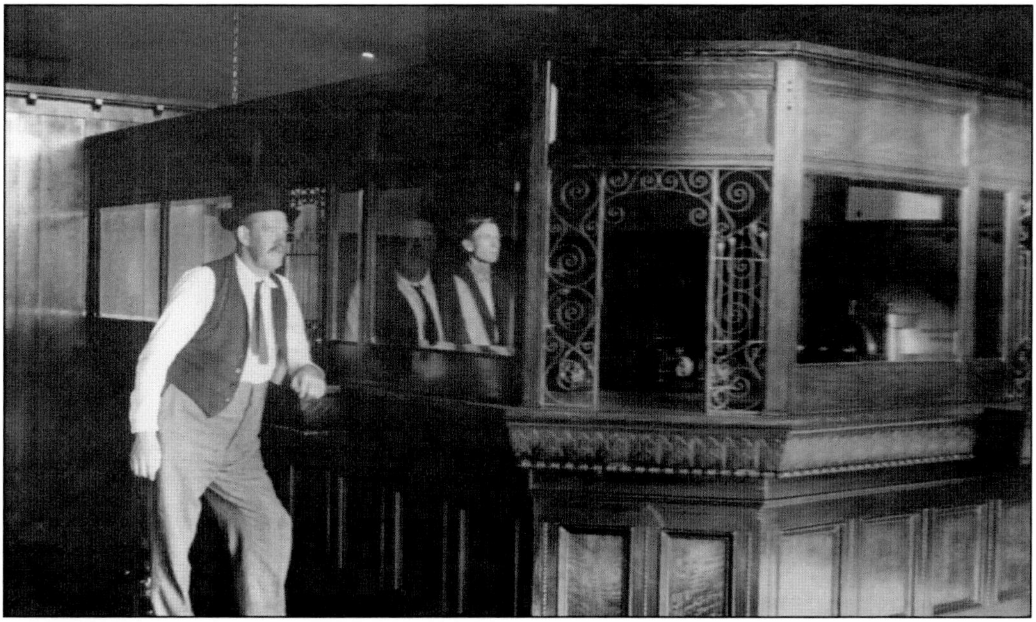

Charles Wolters, building owner, is pictured here in front of the counter. Behind the counter was the bank manager, Edward Adamson, who was living in Minnesota when he saw the advertisement for the open manager position; he then rode the train to Talent, got the job, and returned to Minnesota to bring his family back to Talent. Adamson stayed on until 1927, when the bank's assets were sold to an Ashland bank. (Jerald McGrew.)

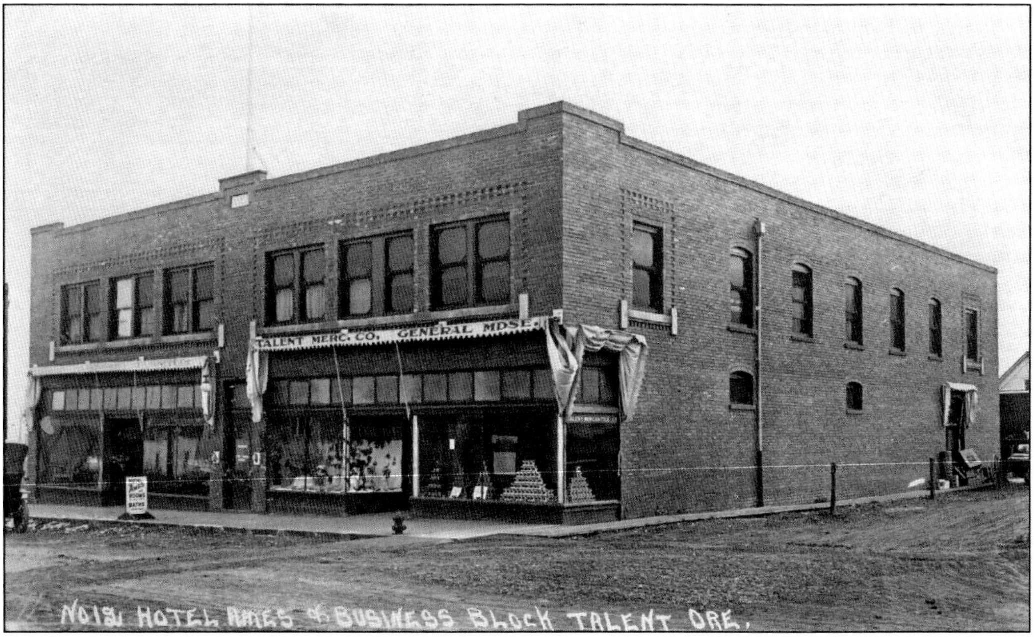

The Ames building was constructed by Albert and Lena Ames in 1914 on the corner of Main and John Streets. The living quarters for the Ames family were located upstairs in addition to a dance hall where Carl Loveland and his band played every Saturday night. The bottom floor had Charles Brown's Talent Drug and Furniture Store and the Talent Mercantile Store run by Louis Brown. (Ken Kantor.)

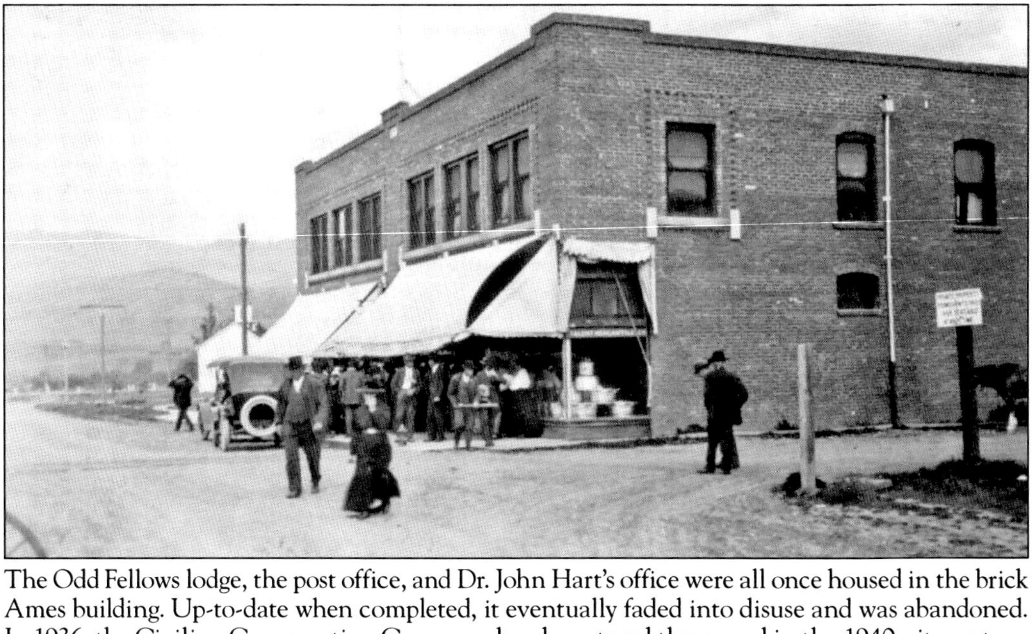

The Odd Fellows lodge, the post office, and Dr. John Hart's office were all once housed in the brick Ames building. Up-to-date when completed, it eventually faded into disuse and was abandoned. In 1936, the Civilian Conservation Corps was headquartered there, and in the 1940s, it was torn down. (Mike Cotta.)

Hanscom Hall, located at 201 Talent Avenue, was constructed in 1906 and is the only downtown building on the National Register of Historic Places. It was a two-story building until this fire damaged the upper floor. The fire occurred sometime between 1913 and 1920. (Jerald McGrew.)

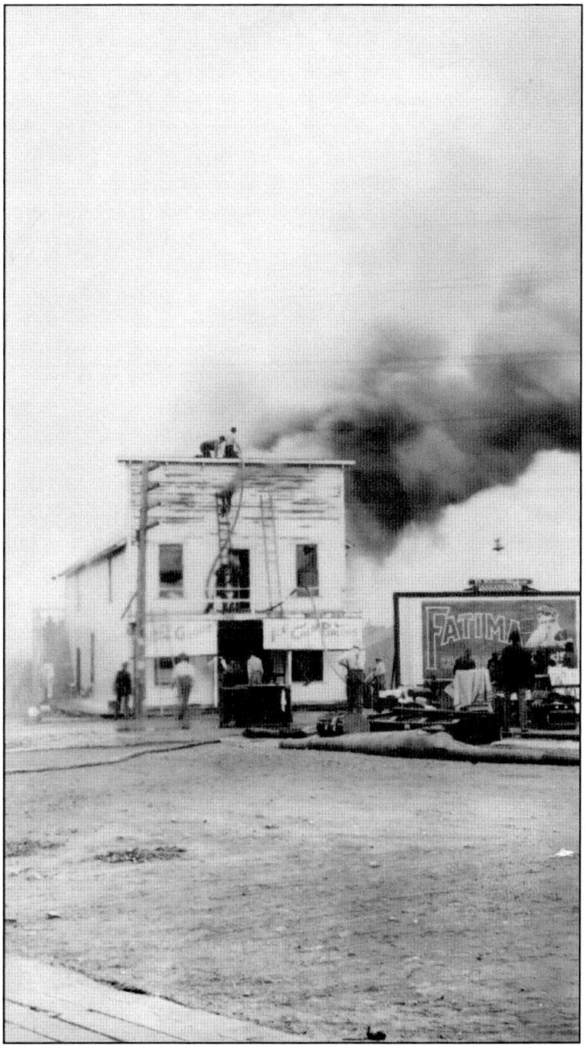

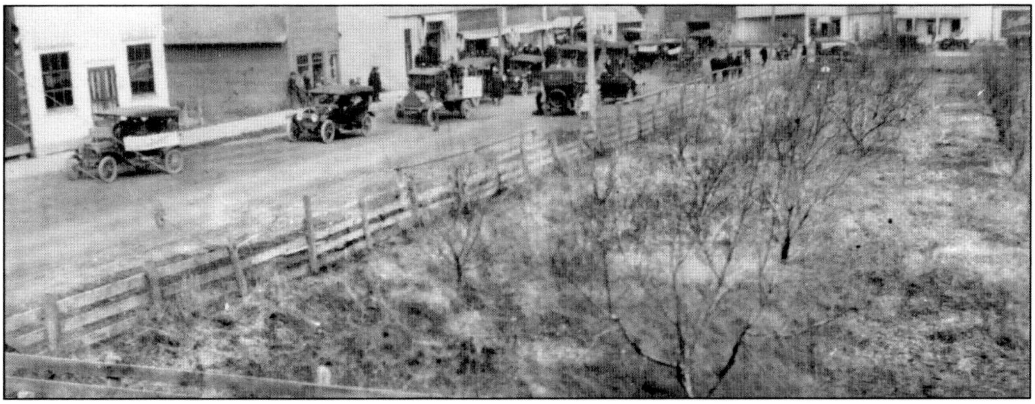

Ashland has Lithia Park, but Talent once had an orchard in the main business district. The orchard was in front of the Wolters home and generated extra income for the family. Patio Village, which offers low-income housing for seniors, now occupies the old orchard site. (Jerald McGrew.)

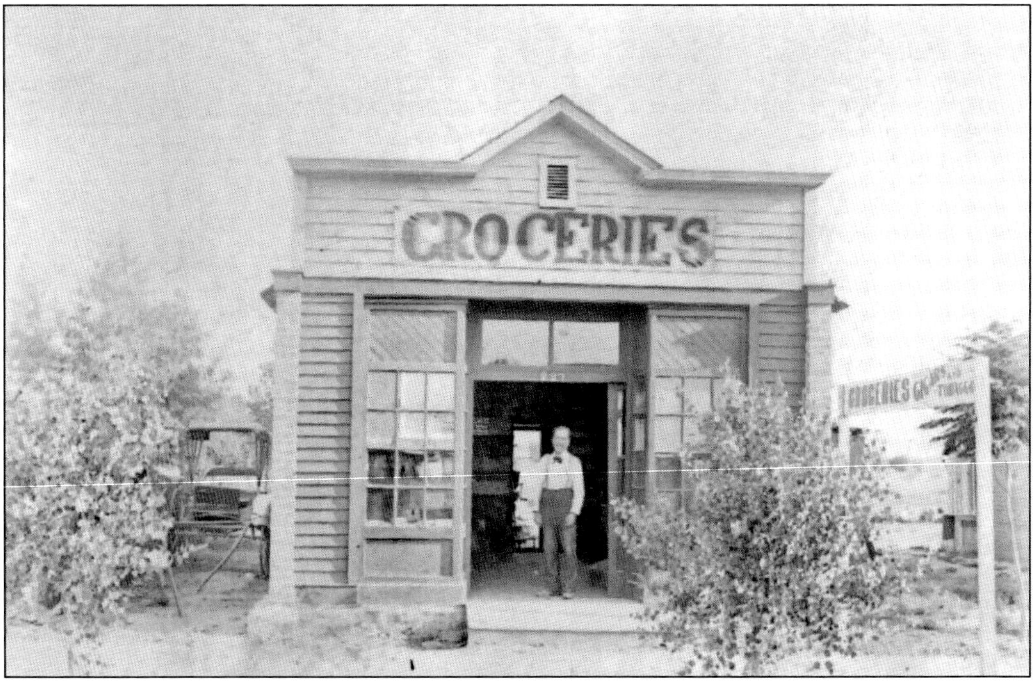

John Van Horn's Groceries was located along Talent Avenue, but little else is known about the store. Granvil Brittsan, Talent mayor in the 1960s and 1970s, was a descendant of John Van Horn. (Sandra Stiefel.)

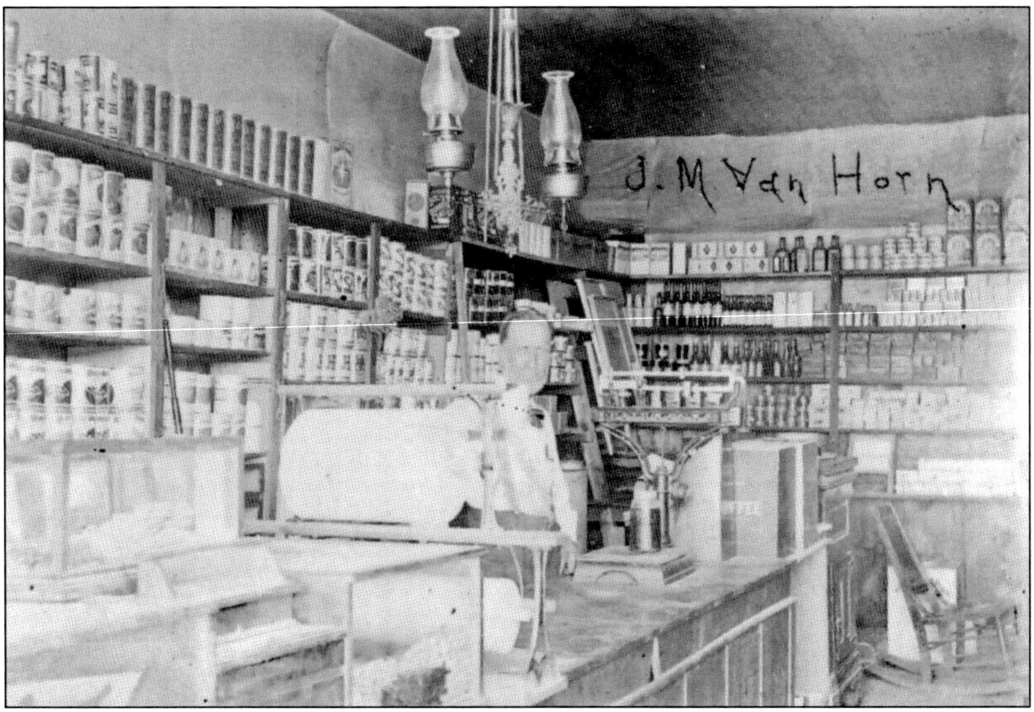

It is unusual to find a photograph of both the exterior and interior of any building. Van Horn's grocery store was small but offered both necessities and luxuries alike. (Sandra Stiefel.)

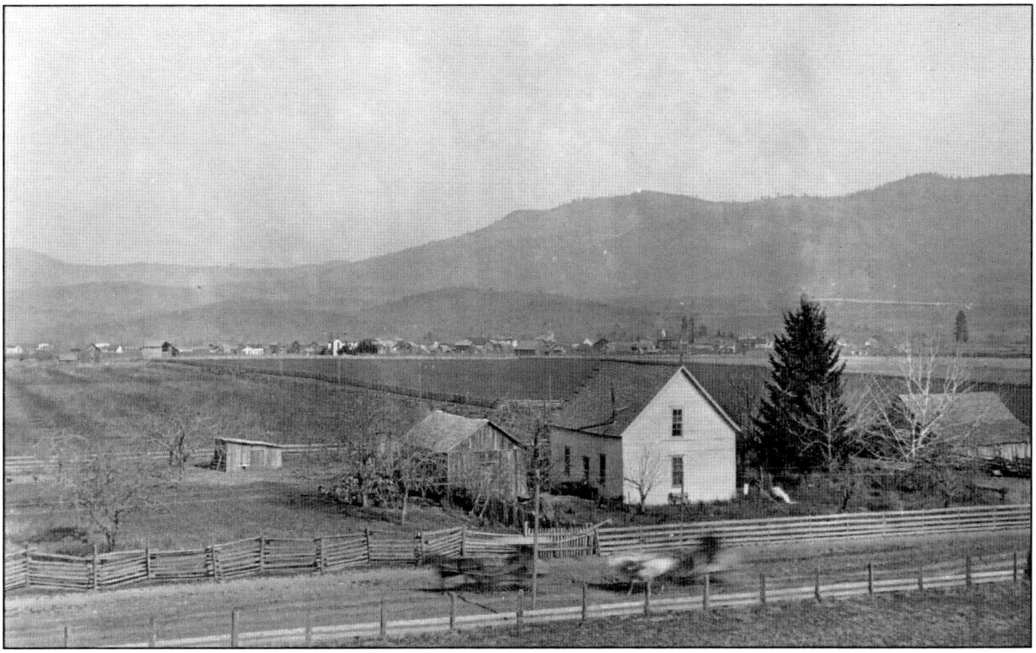

About 1903, two buggies hurried past the William High house, built in 1884, on the corner of Wagner Creek and Rapp Roads. The town of Talent and the road leading to it can be seen in the distance from outside the city limits in what is still informally referred to as Wagner Creek, a distinct district within the Talent area.

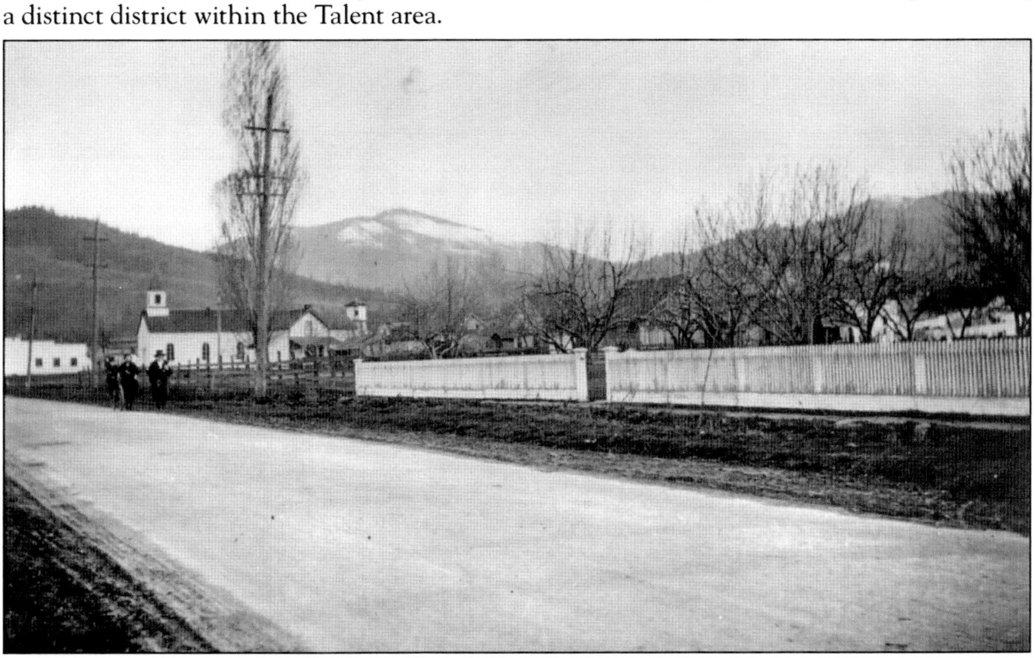

Built about 1871 along Talent Avenue, the Baptist church is said to be the second oldest Baptist church in the state of Oregon. When the Thomas Hill family moved to Talent and wanted to build a store on the church property, they offered to move the building and replace the roof in exchange for the valuable commercial property. The church was moved and has faced Main Street ever since. (SOHS No. 14585.)

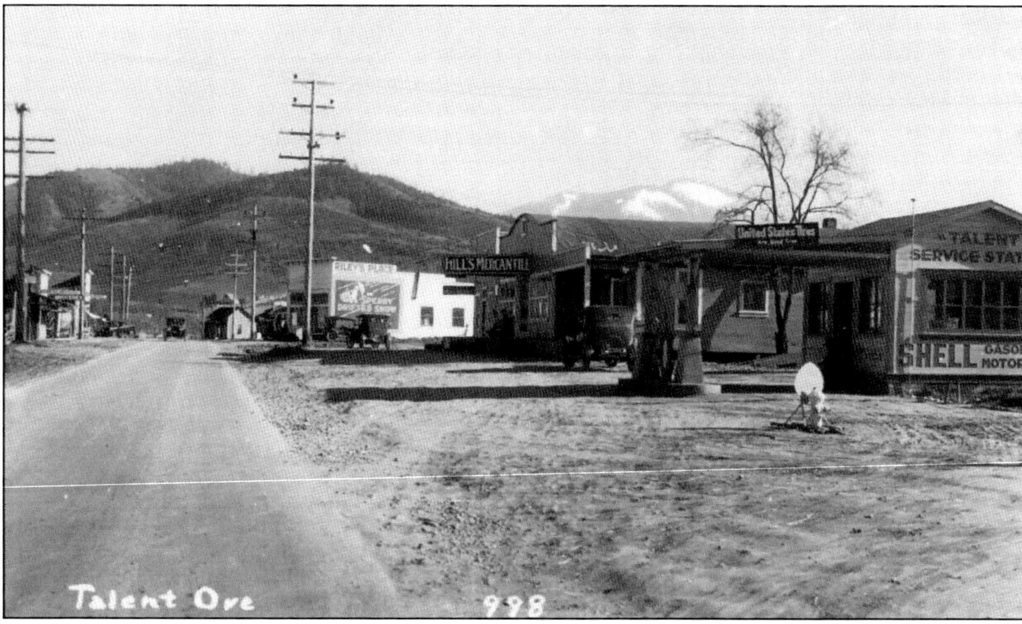

In 1913, Talent became a part of the great Pacific Highway, an ambitious plan to connect Mexico City to British Columbia. The business district was oriented to serve the ensuing automobile traffic and included two service stations, a grocery store, mercantile, blacksmith shop, and meat market. In 1921, Hill's Mercantile was built on the site of the old Baptist church.

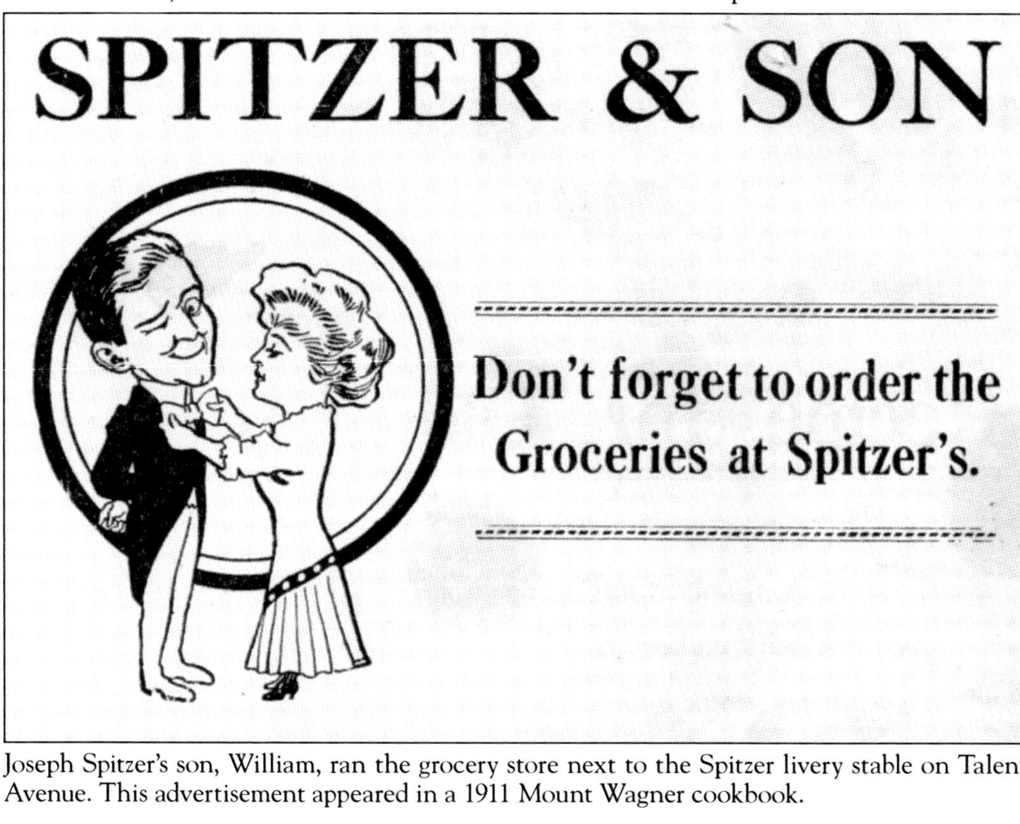

Joseph Spitzer's son, William, ran the grocery store next to the Spitzer livery stable on Talent Avenue. This advertisement appeared in a 1911 Mount Wagner cookbook.

Blacksmith shops gradually converted to garages as automobiles became more common. The exact location of this garage is unknown, but it may have taken the place of the blacksmith shop that was close to the tracks along Wagner Street. (Kerby family.)

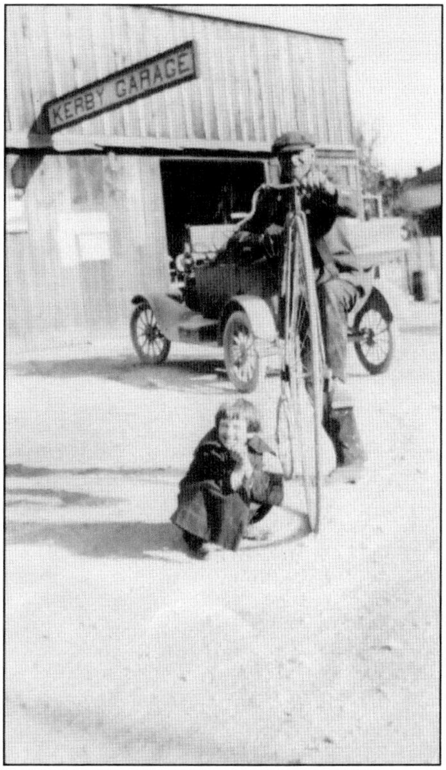

According to the 1911 town directory, Talent Meat Market was owned by Fred Erwin. This photograph, taken after 1911, shows that the name of the market was retained even though it may have changed ownership. Two Gleim relatives came to the market that day to have their picture taken with the butcher and to purchase some fresh goods. (Harry Gleim.)

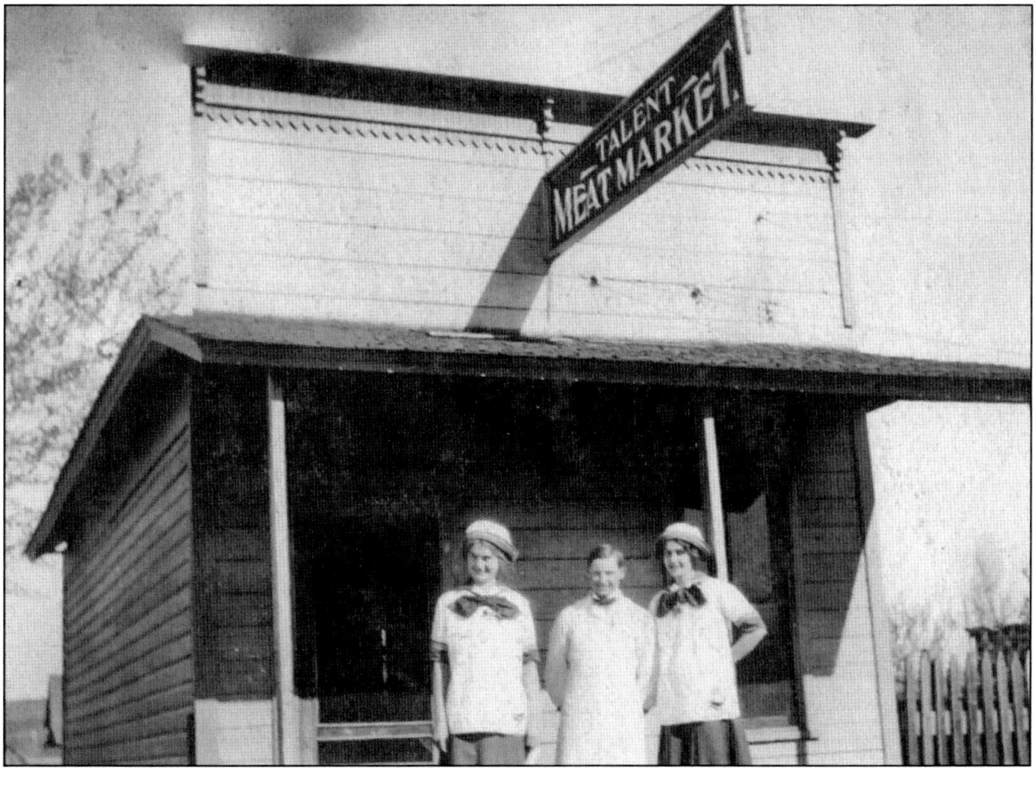

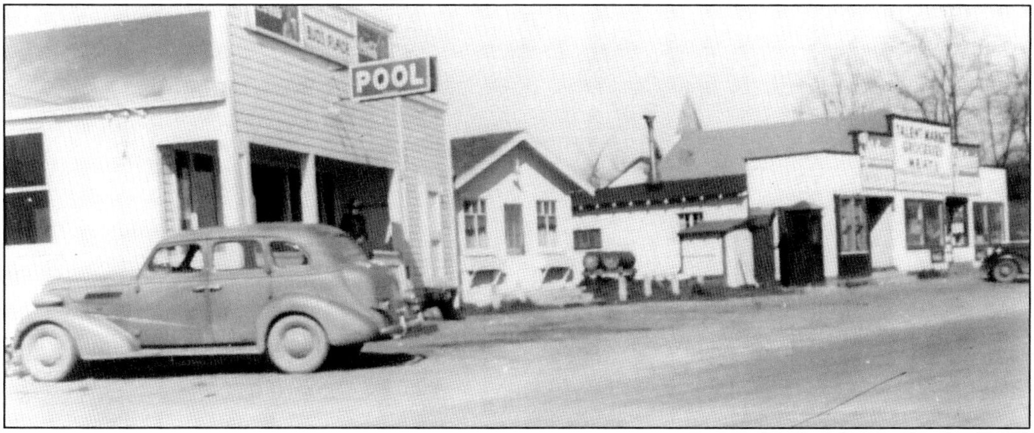

Riley's Place, a pool hall and bar, now the Talent Club, was owned by Riley Nyswaner. He had a sign near the door that read, "If your kids don't have shoes—you get no booze." (SOHS No. 7074.)

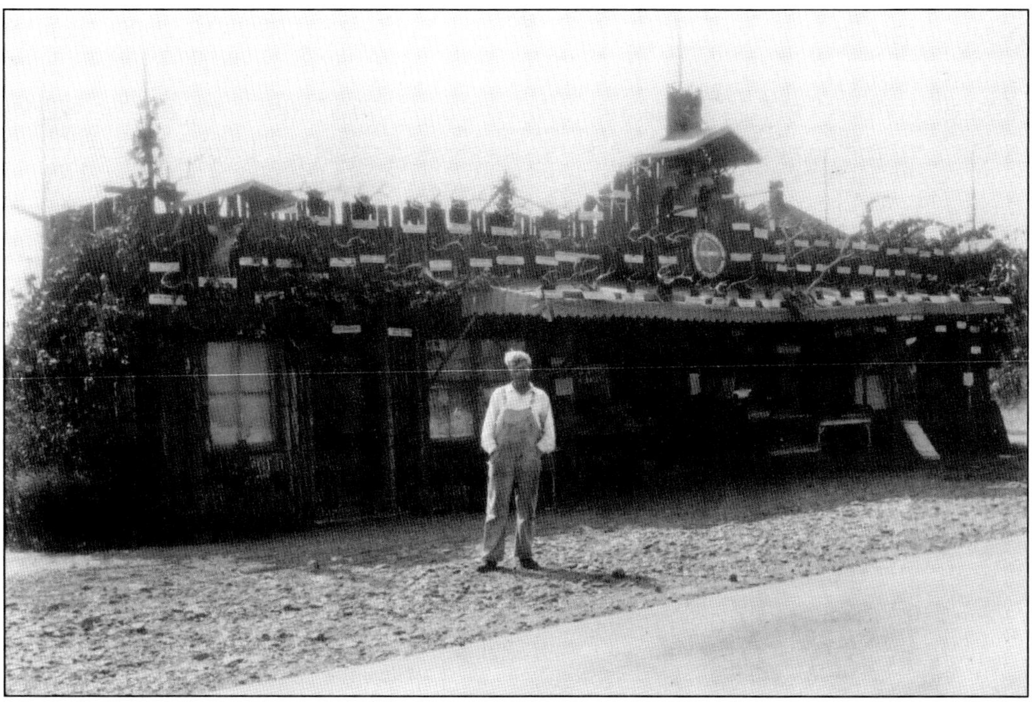

This unusual building would probably not be allowed today with modern construction standards. Earl Simmons owned and operated the produce stand called the Pumpkin Center that stood on the corner of Sunny Street and Talent Avenue. He decorated the outside of the building with deer and elk antlers brought in by local hunters. The building was destroyed by fire around 1960.

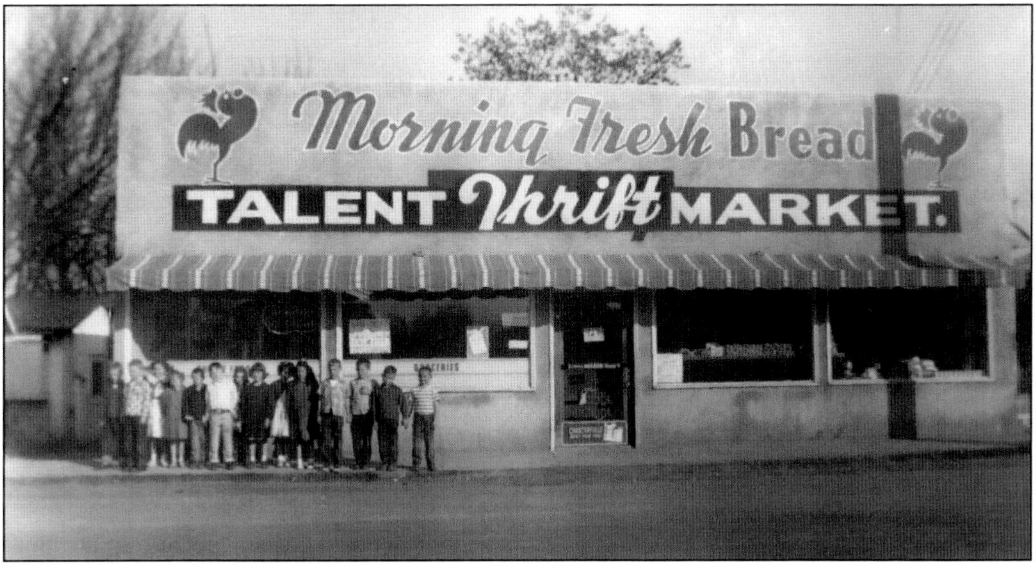

In the 1950s, Talent teacher Genevieve Holdridge took her students all over town to acquaint them with their community. They visited the post office, all the grocery stores, restaurants, and service stations and made a map of where each student lived. The second-grade class is pictured here in front of Talent Market, where Inti Restaurant is located today at 109 Talent Avenue.

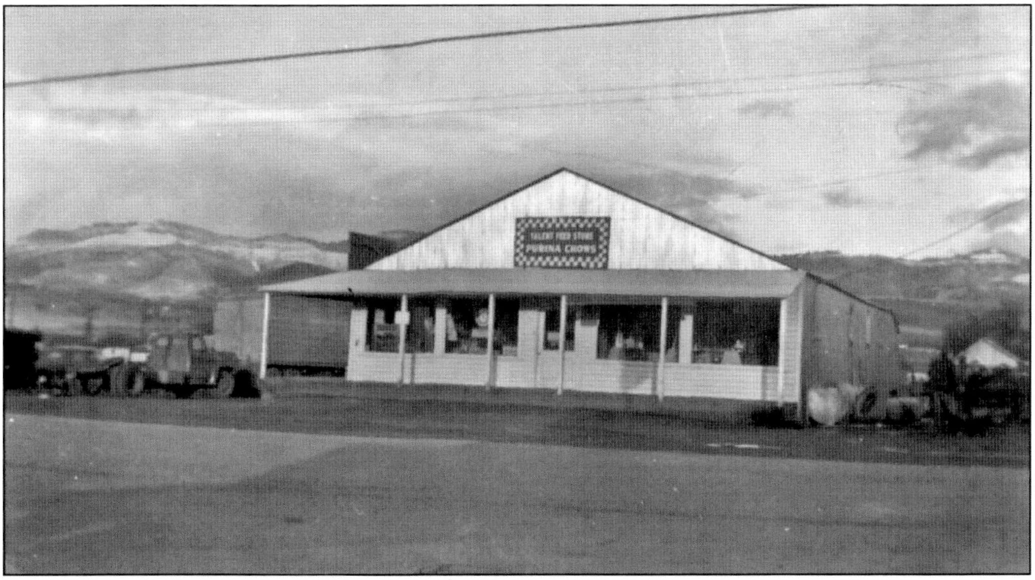

What is now the Camelot Theater was once a feed and seed store called the Talent Feed Store. The Southern Oregon Experiment Station in Talent developed world-famous alfalfa seed and sold it under the Rogue River brand name. It was sent all over the world to help farmers grow pest-resistant and fast-growing alfalfa.

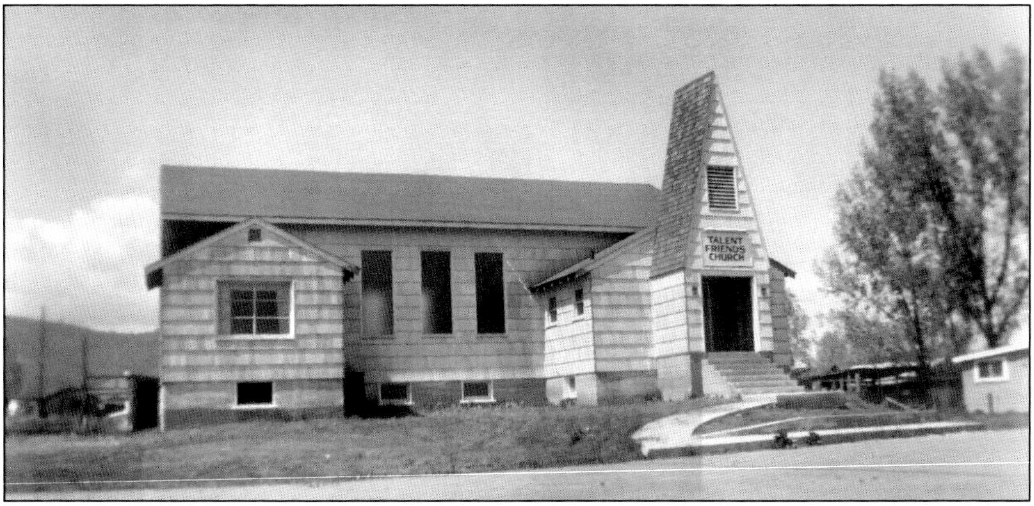

The Quaker or Friends Church has had followers in Talent since the 1940s. They first met in individual homes and in the Baptist church building and then constructed this church house for their people. The front steeple has been removed, but the building on Talent Avenue is still recognizable.

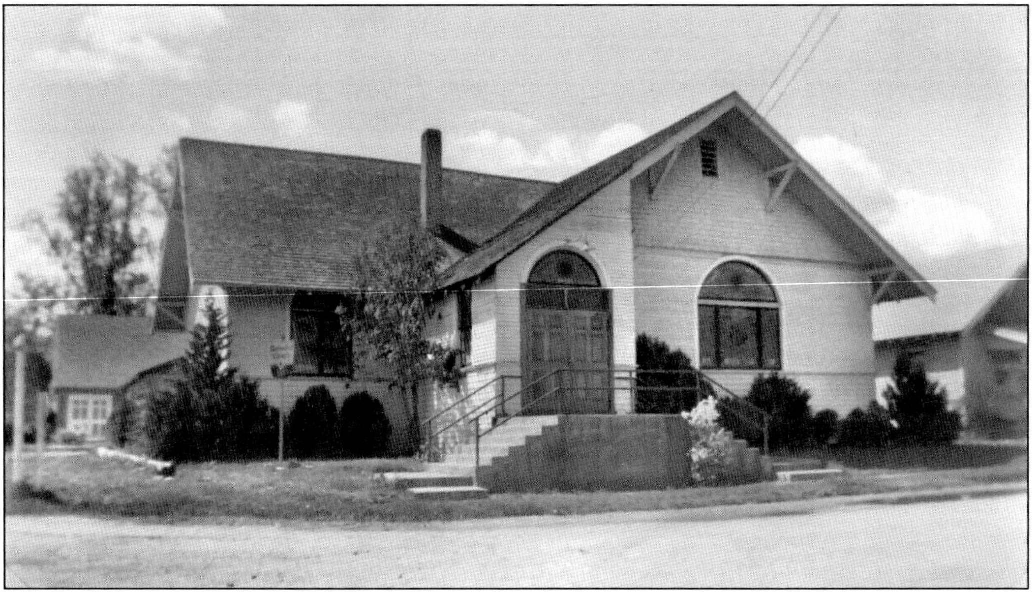

The Methodist Episcopal church was organized in 1909 by a group of 16 people. Before the church house in this photograph was built in 1912, the congregation met in what is now the Community Hall and above the Talent Café in the Hanscom building. Among the many pastors who have served the church, Alice May Woolley, who served in the 1950s and 1960s, is fondly remembered for ministering without controversy, though female pastors were uncommon.

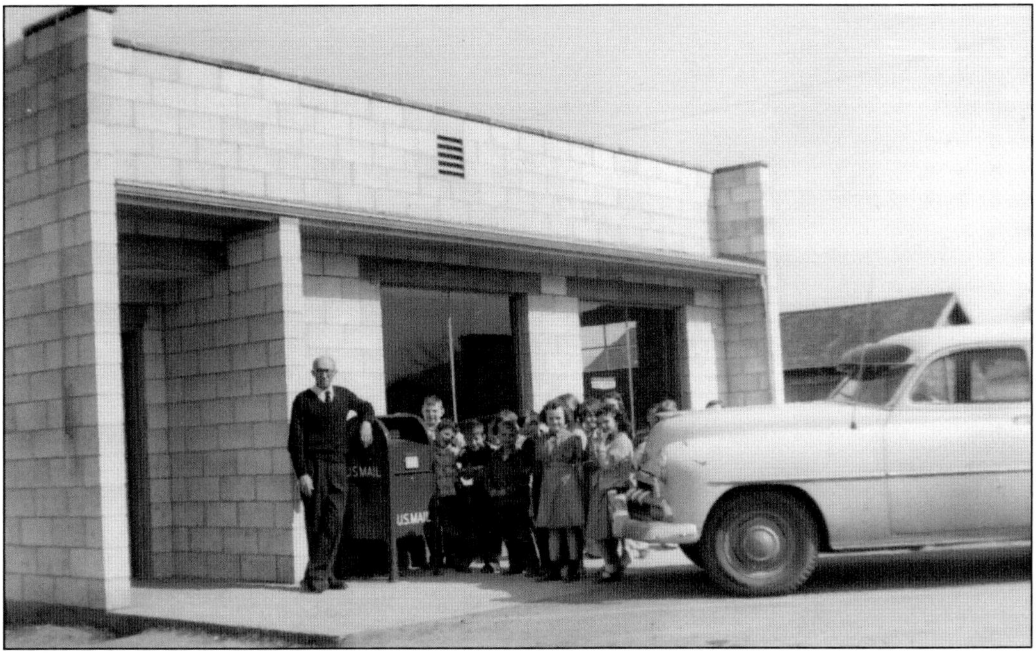

The post office has been housed in several buildings in Talent. Charles M. Harvey was the first postmaster in 1883 but was replaced months later by Aaron P. Talent. Mr. Smoot, the postmaster in this 1958 photograph, showed the second-grade Talent Elementary class the new post office. The building is now the Suds 'N' Duds Laundromat and is located on Market Street.

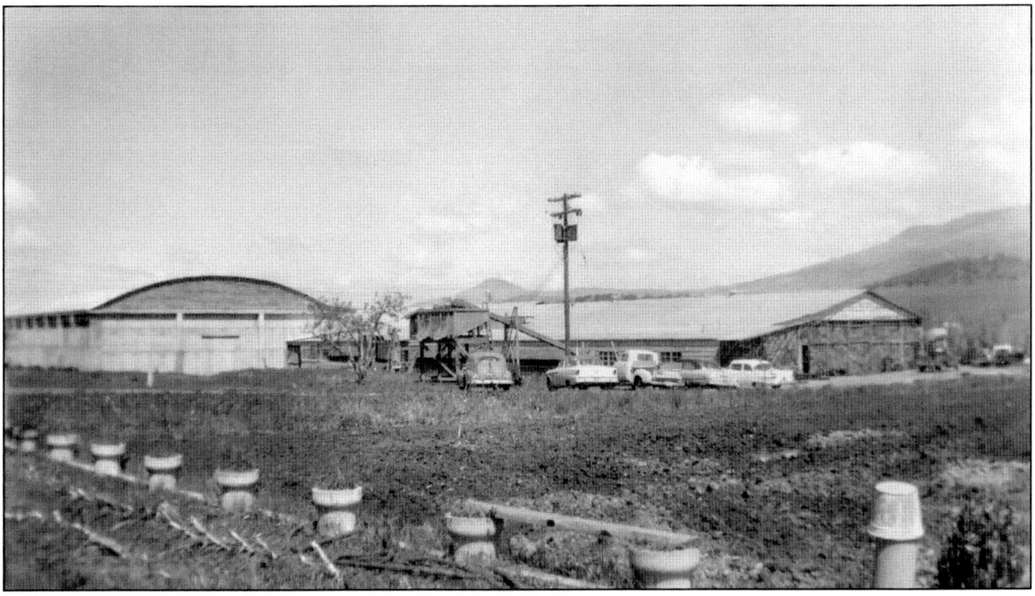

Barnes Packing House stands on the corner of Suncrest Road and Highway 99 as a reminder of how important the fruit industry has been to the Talent area. The empty boxes stacked in the back of the building now were once loaded with fruit in the harvest season.

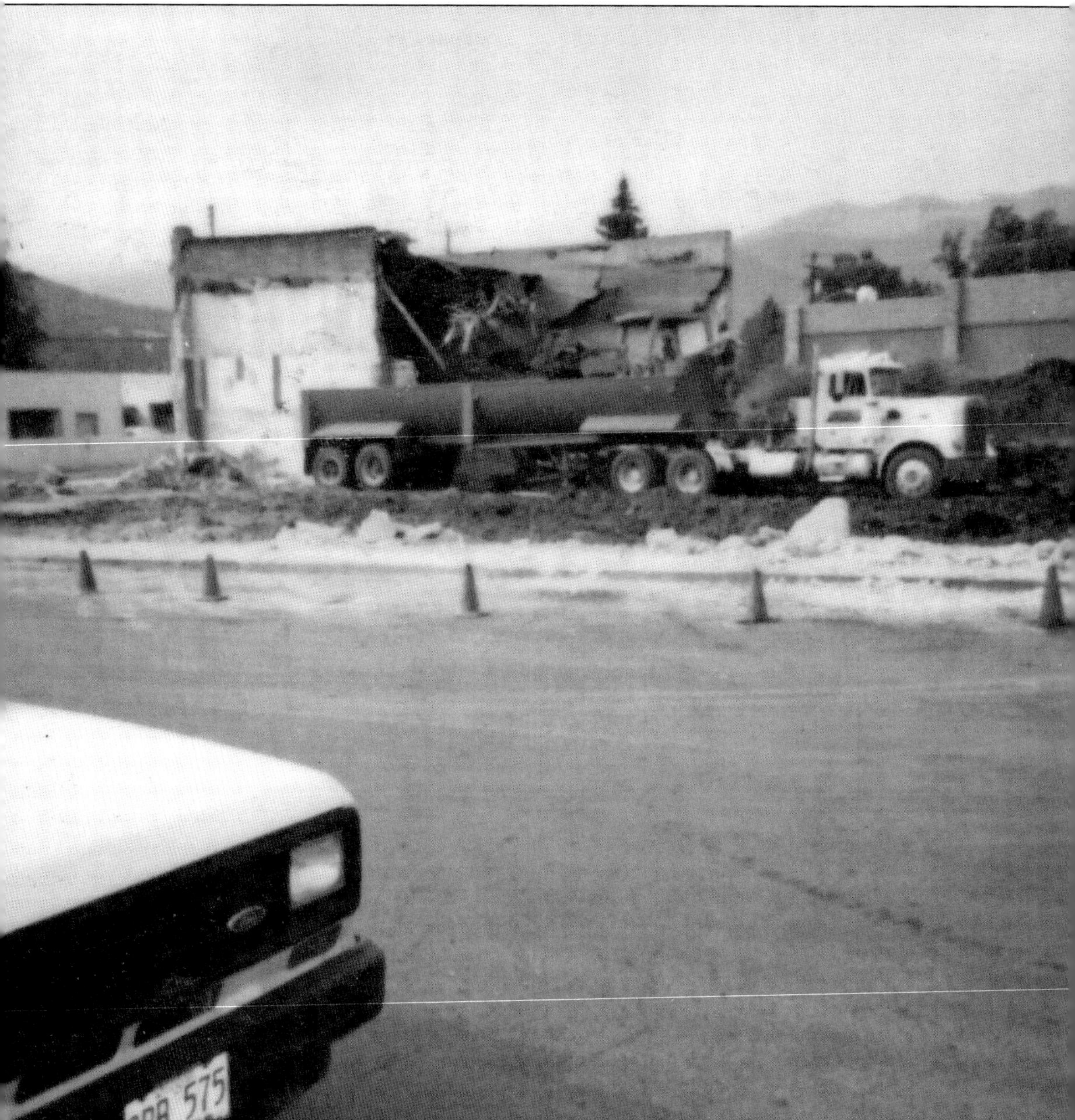

The old Wolters building that served Talent for 80 years was finally torn down in 1991. A landmark building, it had many owners and uses over the years. When a new Rick's Market was built, the space was needed for a parking lot. (Ron Ridgway.)

Five

OCCUPATIONS

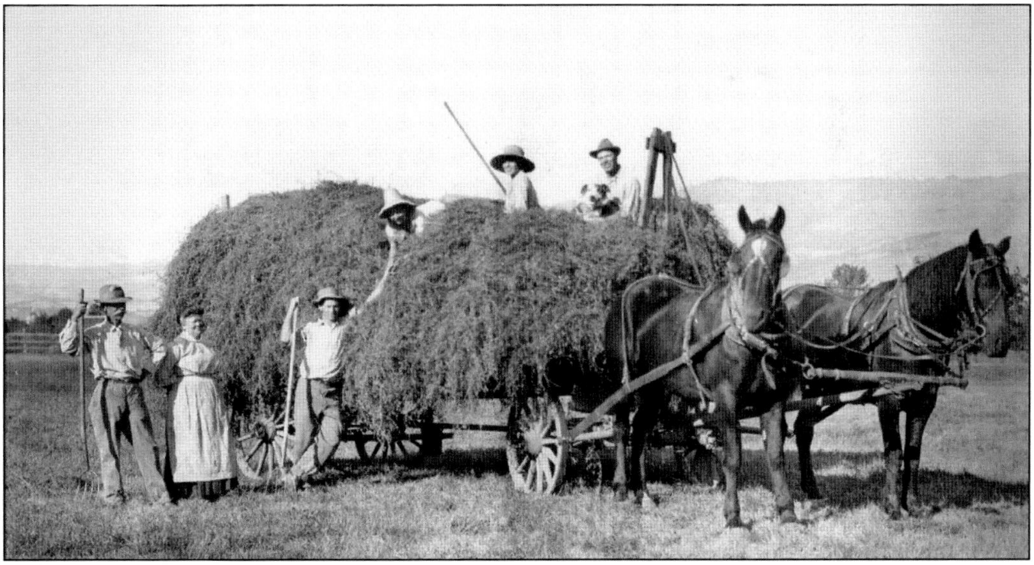

From left to right are Harry S. and Frances Gleim, Harry N. and Louella Madge Gleim, and Cora and Will Gleim, who worked together during the hay harvest. Harry and Madge Gleim, who were married in 1924, had just enough time to touch hands as the photograph was being taken. (Harry Gleim.)

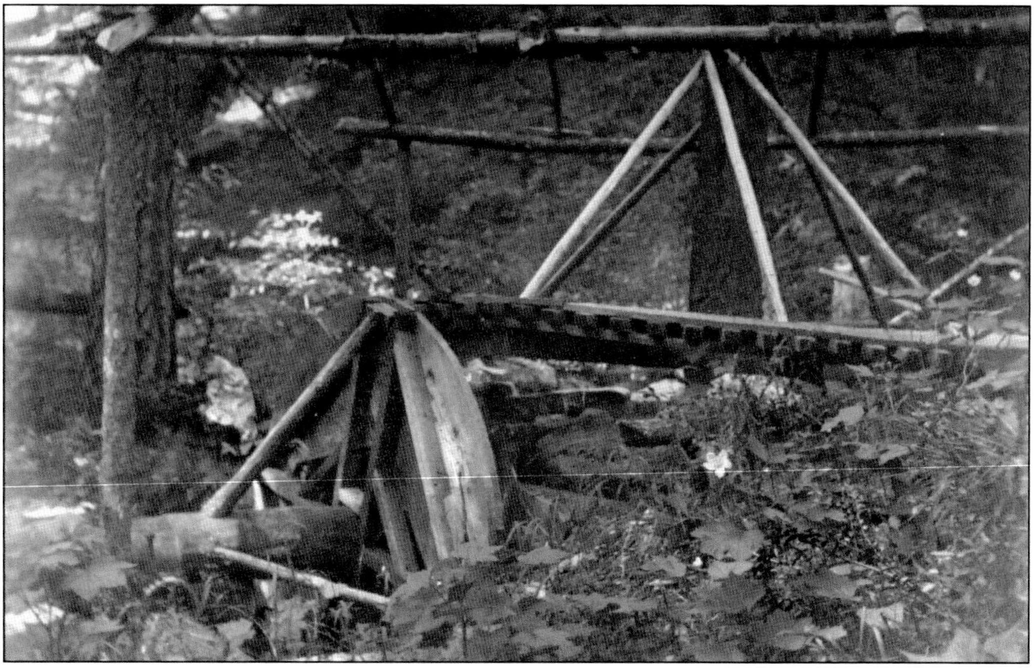

Someone went to a lot of trouble to make this device, but it is unclear what it may have been used for. If the wheel, located up Wagner Creek, was made of stone, it might have been an arrasta used to crush rock to extract minerals. (Kerby family.)

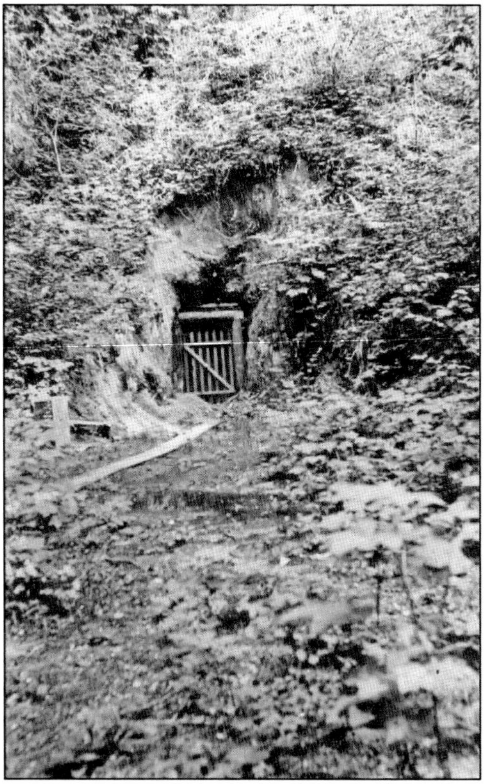

Many of the early settlers had been miners who never quite gave up the dream of discovery. They kept digging even when the expense of extraction exceeded the rewards. This mine shaft was in the Wagner Creek canyon. (Kerby family.)

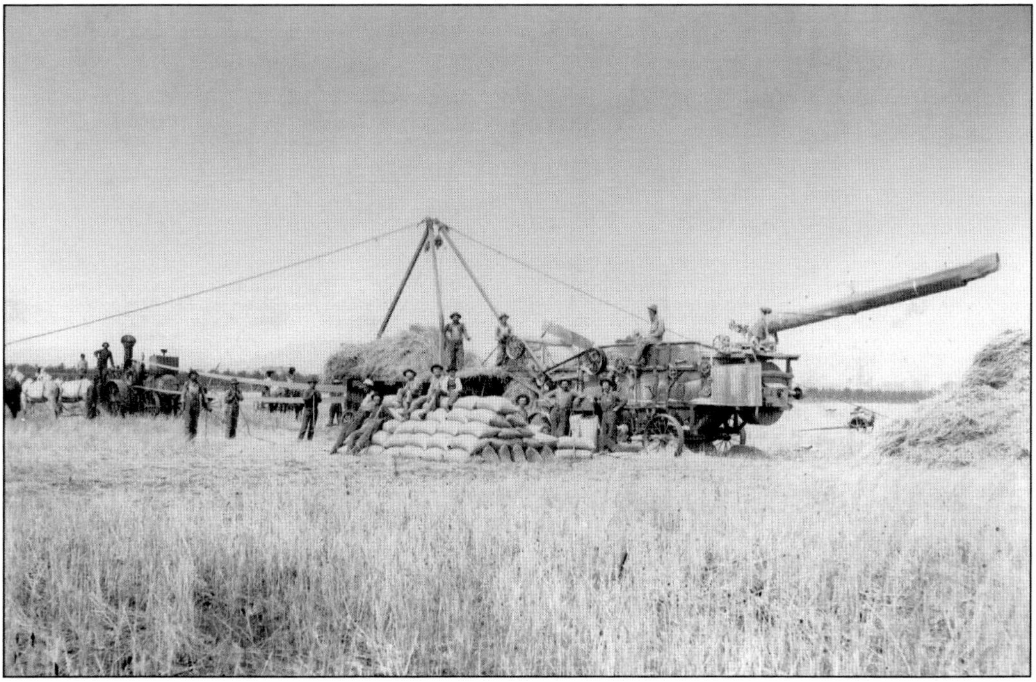

This elaborate machine was owned by Joshua Patterson and was designed to thrash wheat and bag it. Belts driven by steam power made the machinery work after men fed the machines with the harvested materials. (Richard Allen.)

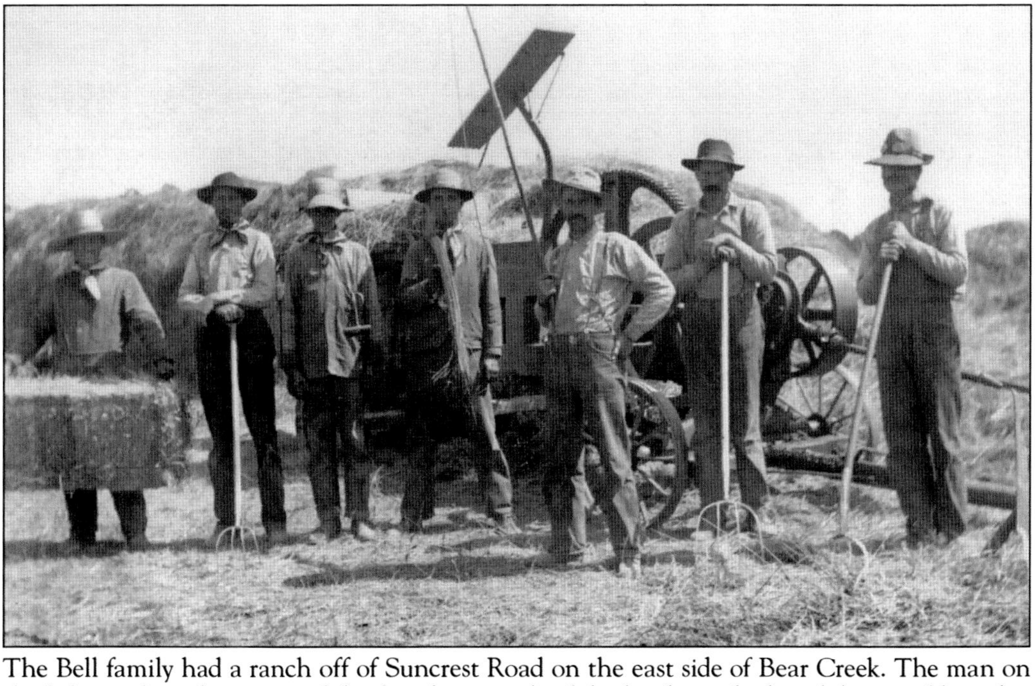

The Bell family had a ranch off of Suncrest Road on the east side of Bear Creek. The man on the left may have been one of the first from south of the border to find work here and have his picture taken. Currently much of the agricultural work in the valley is done by Latino men and women. Their presence is a vital part of the Talent community.

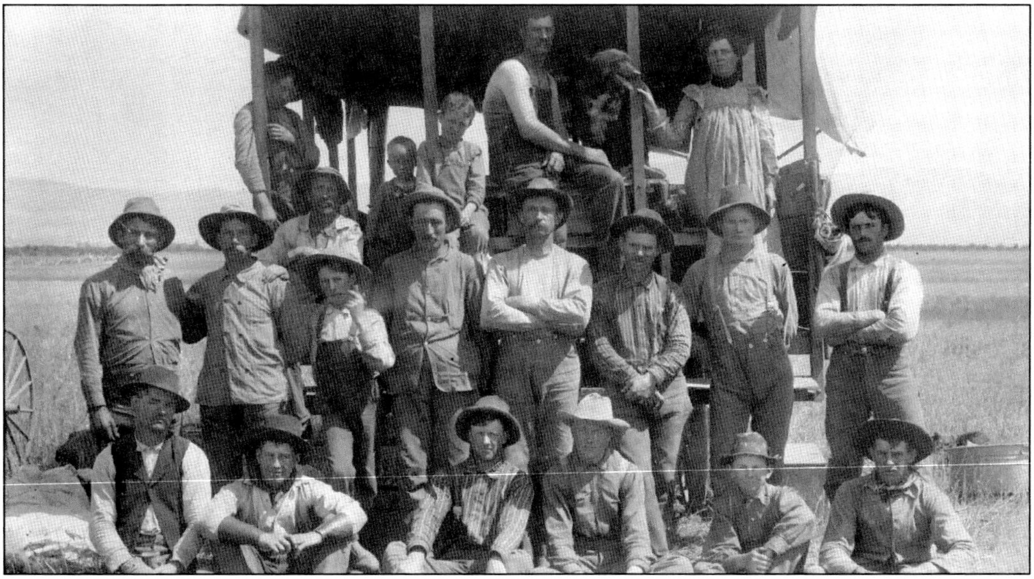

No matter how skilled a farmer was, he could not make a success of his enterprise without hired hands or cooperative neighbors. The woman who delivered the noon meal in this image allowed her picture to be taken in spite of her obvious pregnancy. One of the youngest workers (third from left, second row) tried to look older by posing with a pipe in his mouth. (Jerry Kime.)

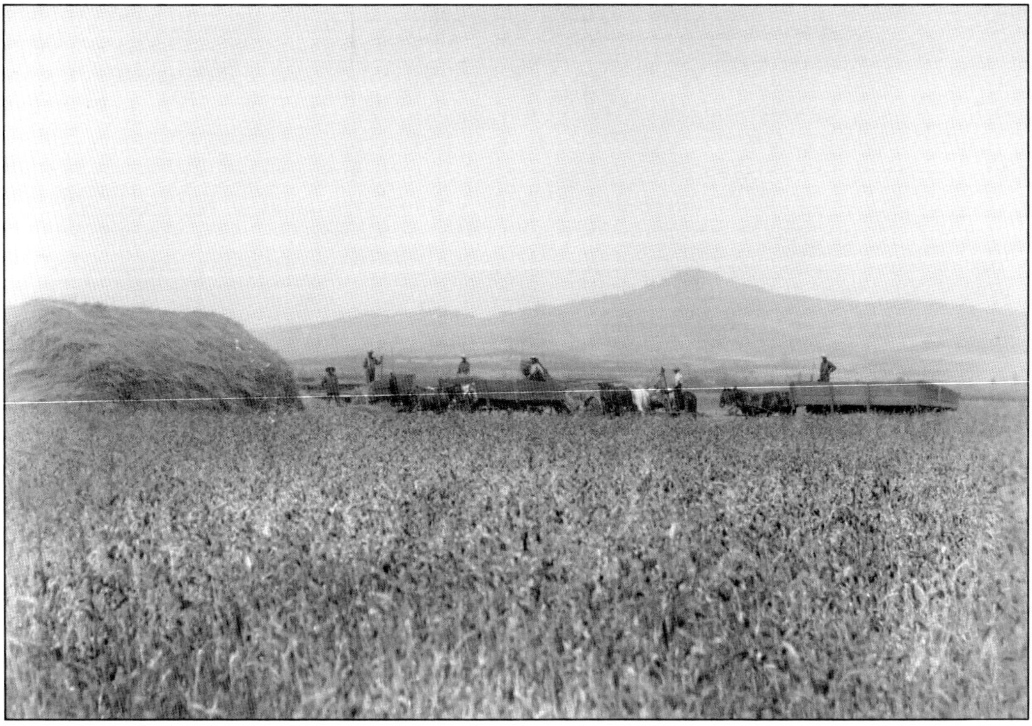

The unmistakable landmark Roxy Ann Peak, in the Cascade Range, identifies the approximate spot as this field near Bear Creek. Talent was first and foremost an agricultural settlement, and farmers outnumbered all other occupations. Today Talent is more of a place to live than a place to work. (Jerry Kime.)

George Kerby moved tons of hay from the fields to the barn over and over again. Hay grew lustily up in the hills and supplemented livestock feed through the winters. (Kerby family.)

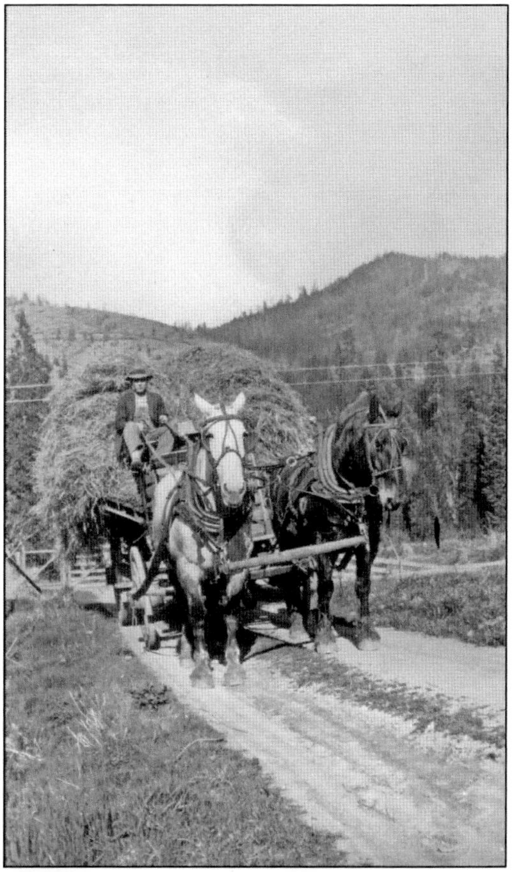

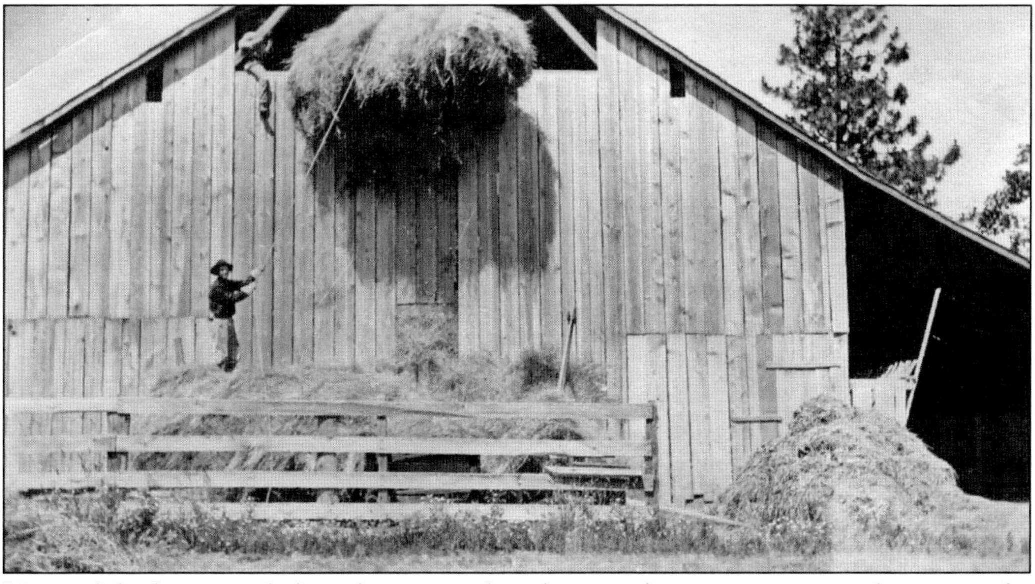

Many of the barns, including the one in this photograph, are gone now, making room for development. This particular barn was built by the Beeson family and was a part of their everyday life for decades. Will Gleim was putting hay in the barn for winter use. (Harry Gleim.)

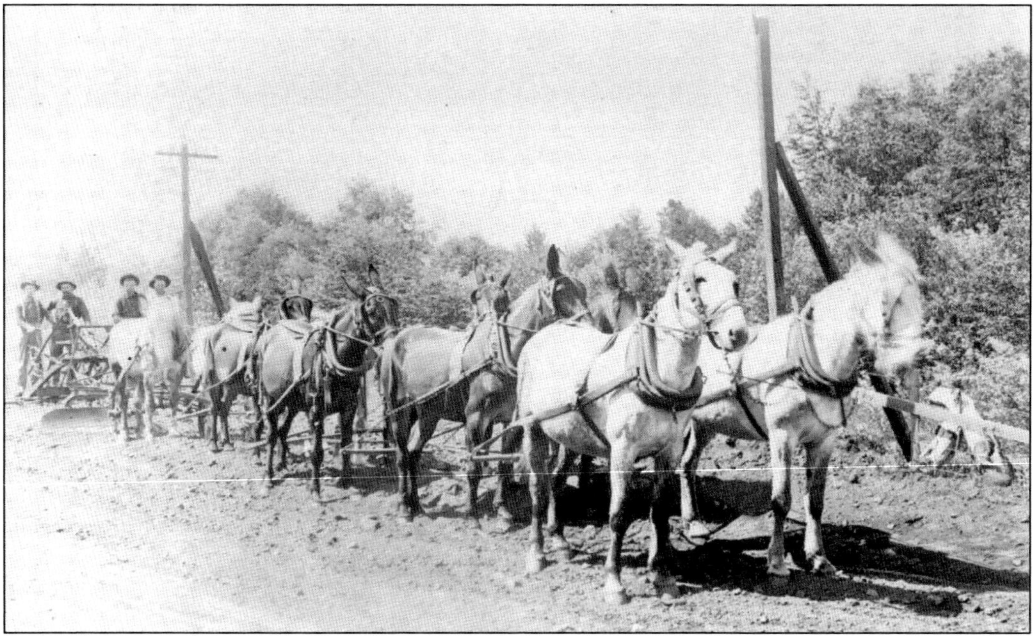

The horse-drawn road grader was a back saver for the neighborhood crews who were used to digging the roads with picks and shovels. The Patterson family had both a road grader and a harvesting machine. They could operate both at the same time if enough horses and men were available. (Richard Allen.)

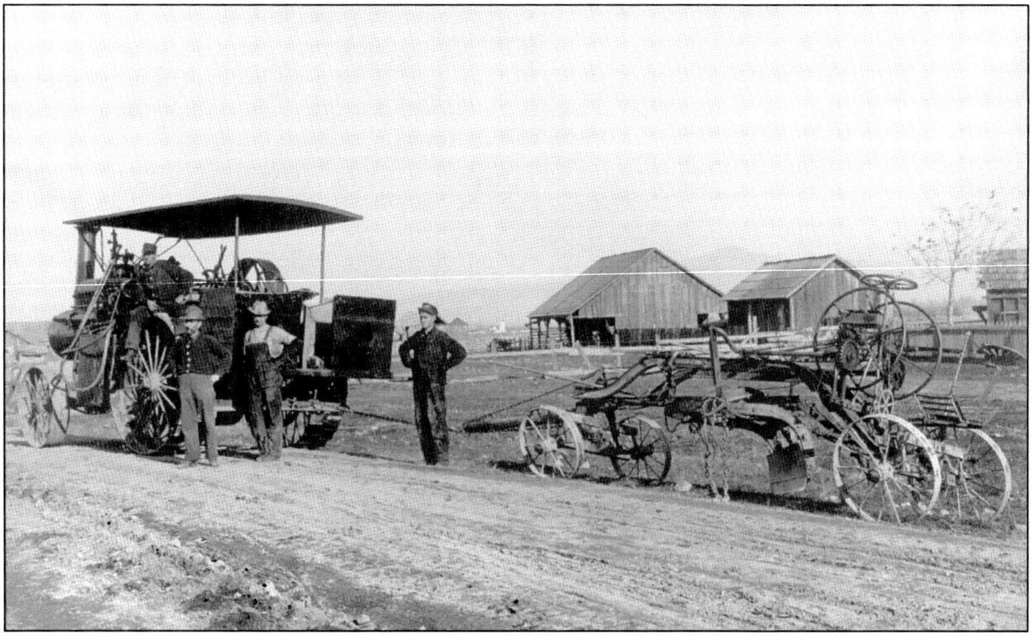

Nicholas Kime (standing on the left in front of the machine) was hired by the county to maintain the county roads with modern equipment. The road grader was powered by the steam engine in front of it. (Jerry Kime.)

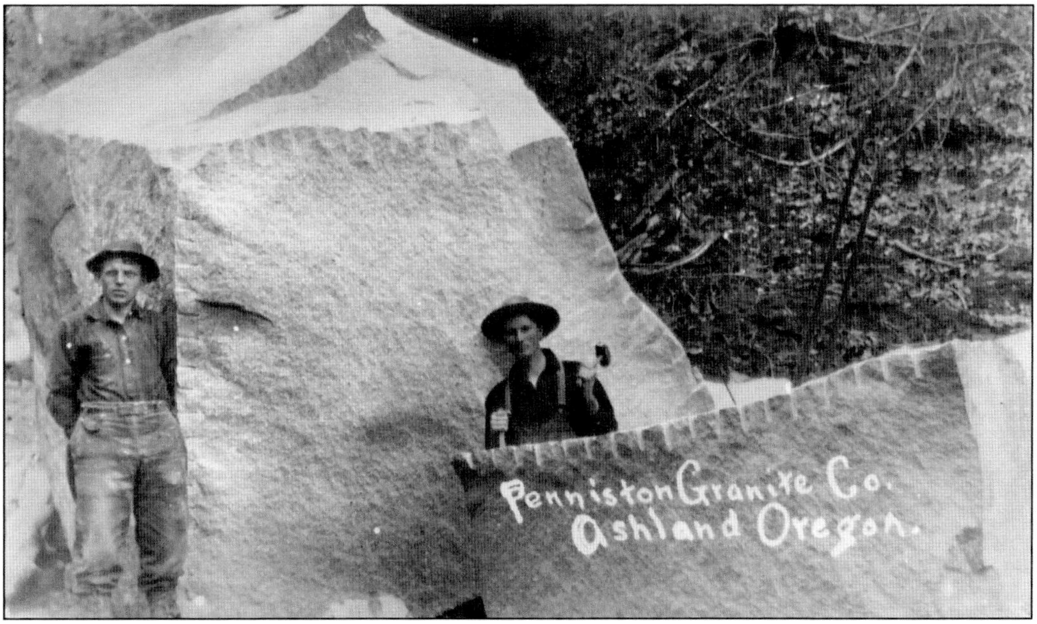

The Penniston Granite Company operated on Granite Street in Ashland, Oregon. The company took out huge chunks of granite to make walls, fences, tombstones, and other projects. With the advent of the automobile, people from Talent could commute to Ashland to work. George Kerby from Talent married the daughter of Samuel Penniston, who owned the Granite Company. (Kerby family.)

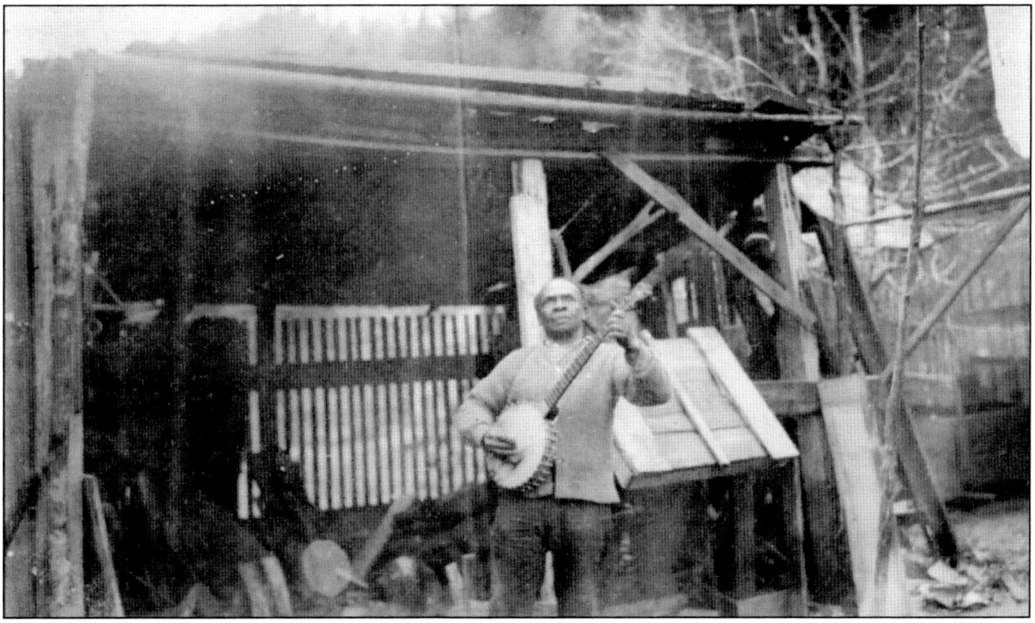

Southern Oregon cannot pretend to be racially diverse, but there are and have been people of color who lived here. The man in this postcard was identified as "Old Black Joe." The back of the card had the following inscription: "Upper Granite Street, Ashland, Oregon, His spirit has gone on." It is likely that Joe worked for or lived near the Samuel Penniston Granite quarry on Granite Street, as his photograph was among that family's papers. (Kerby family.)

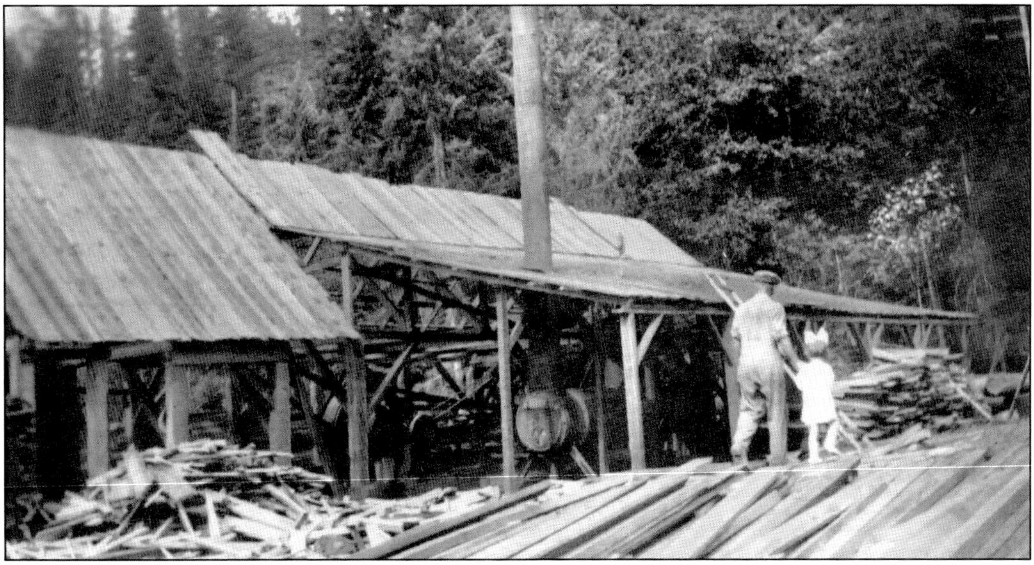

Since 1854, there has been a sawmill on Wagner Creek. Each time a new owner took over the mill, it was moved a little farther up the canyon. In this photograph, the mill was called the Gleim and McGrew Mill and Will Gleim was showing his daughter, Alice, around the work site. (Harry Gleim.)

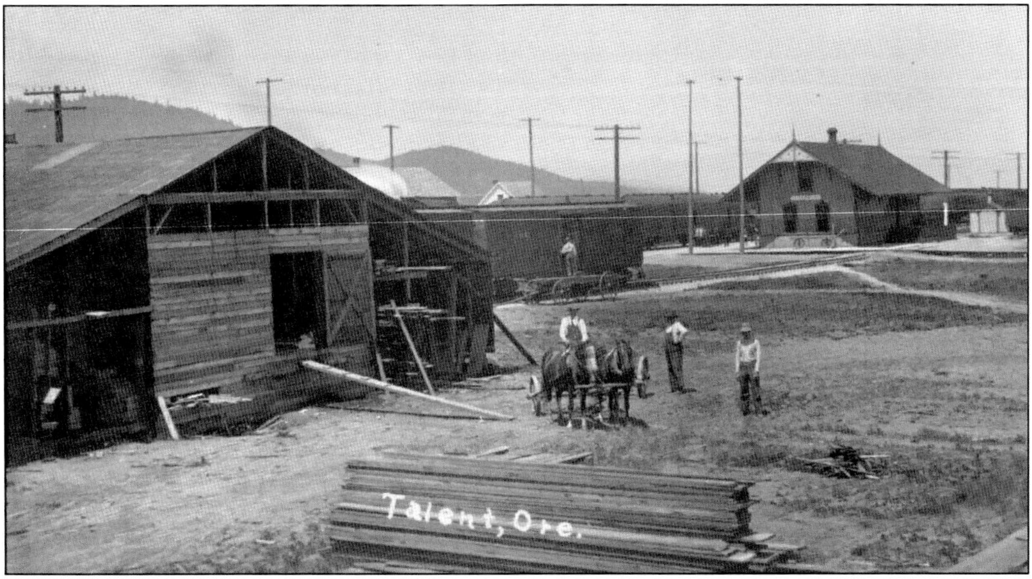

Skateboarders would be surprised to know that there was once a planing mill on the skate park site. A box factory was also in operation in Talent to make the fruit boxes that were shipped to market from the rails of the side track. (Mike Cotta.)

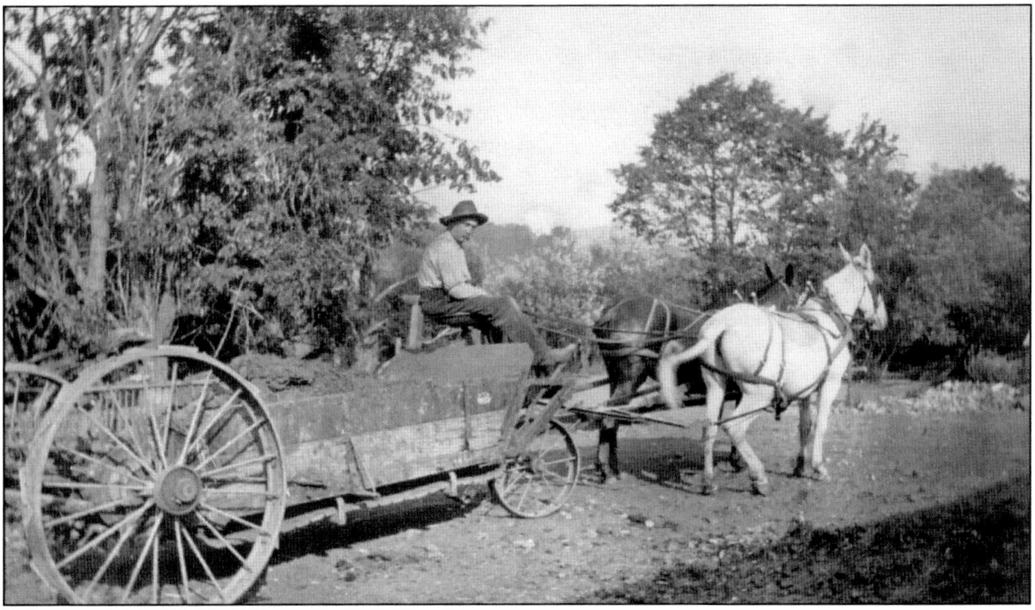

If one has farm animals, a manure spreader is a necessity. Joshua Patterson Jr. is seen in this image spreading manure around to fertilize the crops. (Richard Allen.)

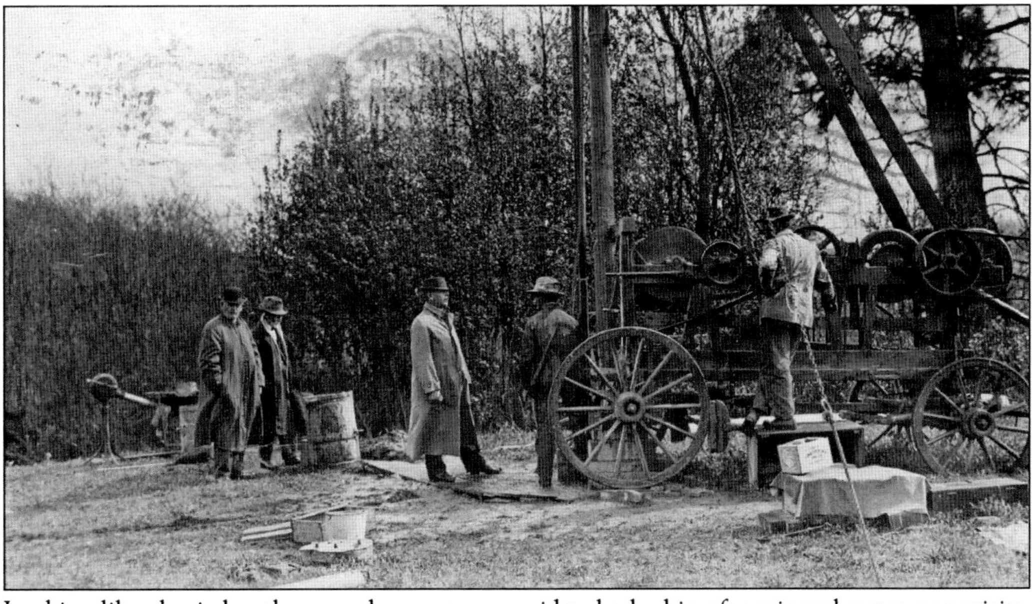

Looking like classic bootleggers, these men are said to be looking for mineral water, not spirits. One developer after another tried to tap into nature's resources to make a quick buck. They were born too soon and today would marvel at the popularity of plain bottled water. (Kerby family.)

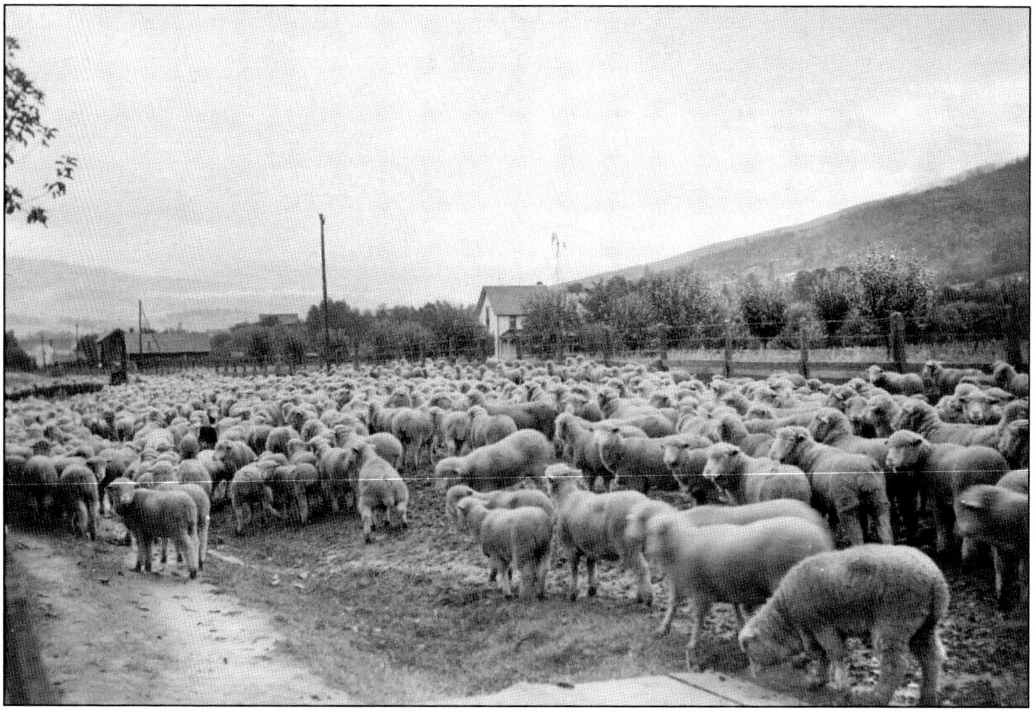

Sheep and cattle found ample grange lands outside of town. Along Rapp Road, these sheep seem to be waiting to be sheared or to be driven up to the mountains to graze. Because the Ashland Woolen Mill burned down in 1900, it would have been necessary after that to ship wool to Pendleton or Salem, where woolen mills were operating.

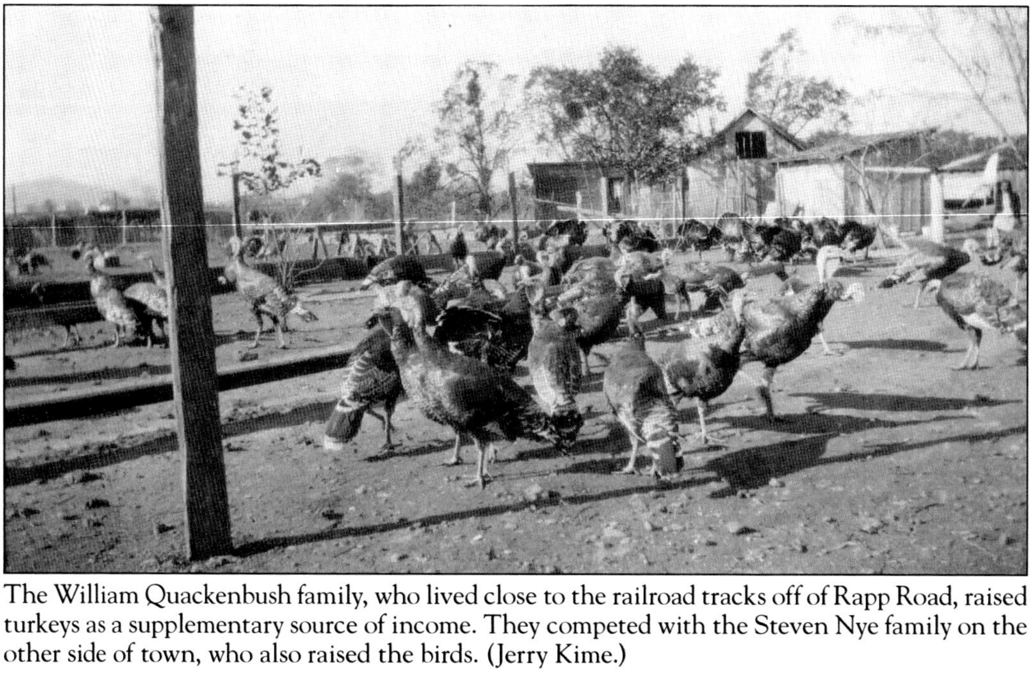

The William Quackenbush family, who lived close to the railroad tracks off of Rapp Road, raised turkeys as a supplementary source of income. They competed with the Steven Nye family on the other side of town, who also raised the birds. (Jerry Kime.)

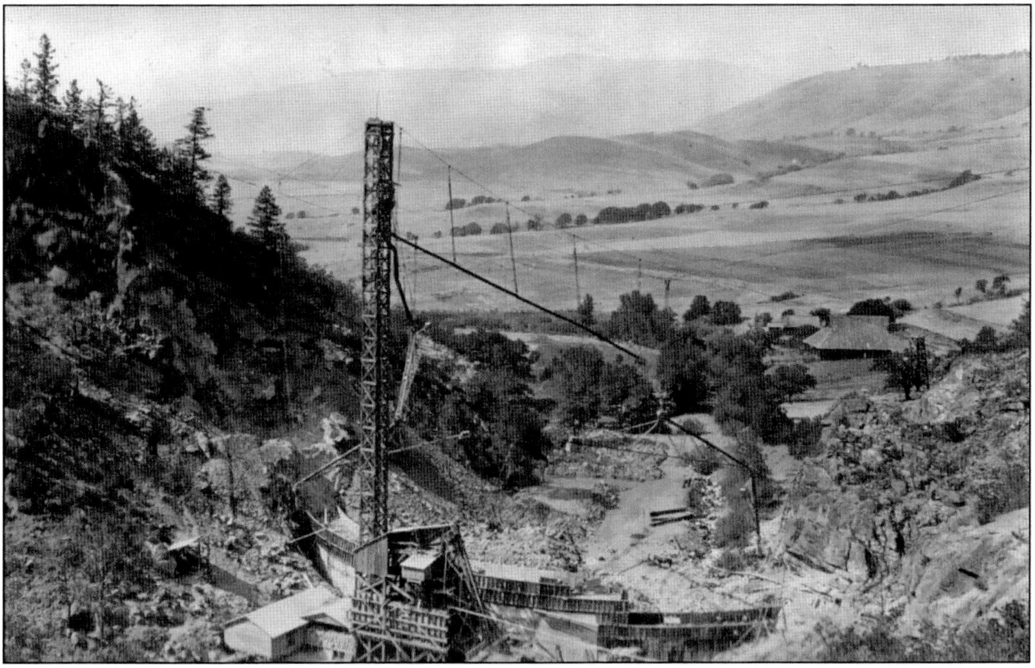

Construction on Emigrant Dam on Route 66 southeast of Ashland began in 1924. It became a major part of the Talent Irrigation District. The lake created by the dam is purely recreational and covers the old Isaac Hill donation land claim (seen on the right).

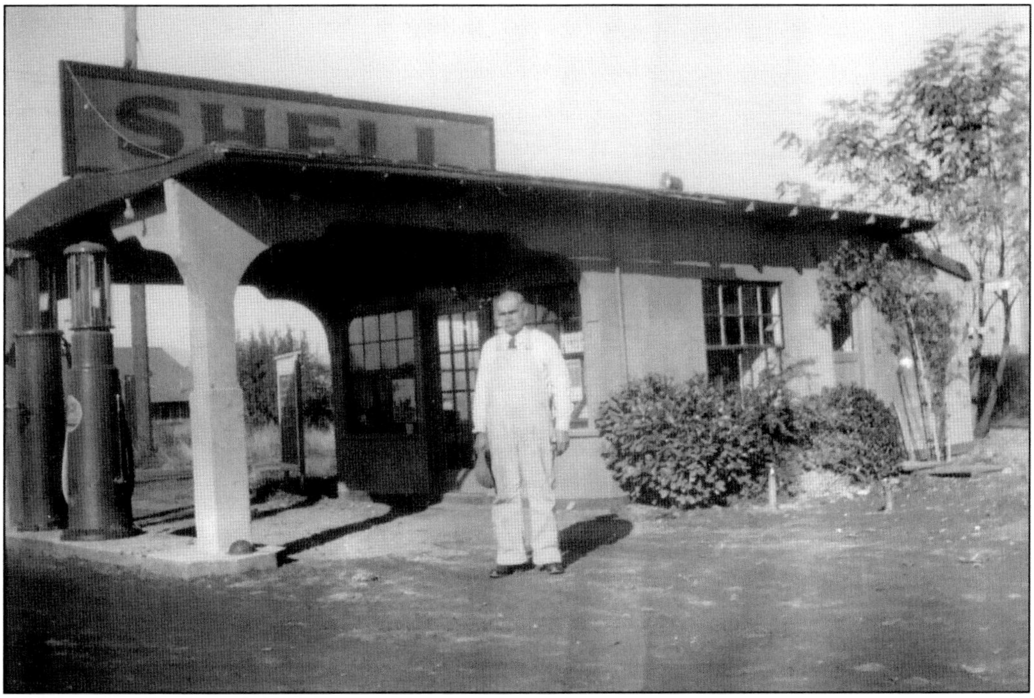

William Hewitt owned and operated this Talent Shell Station on the corner of Suncrest and Highway 99, which was cleverly located to catch traffic from Colver Road, Highway 99, and Talent Avenue.

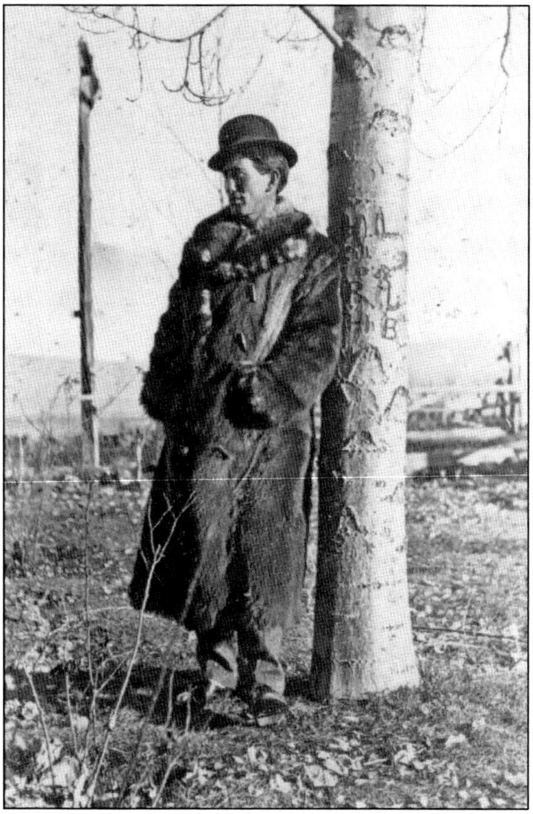

Thomas Newton from Gainesville, Texas, seen here in his big fur coat, was the proprietor of the service station (shown below) when it was called Newton's. He apparently did not stay long and moved to Klamath Falls, Oregon, where he ran a piano store.

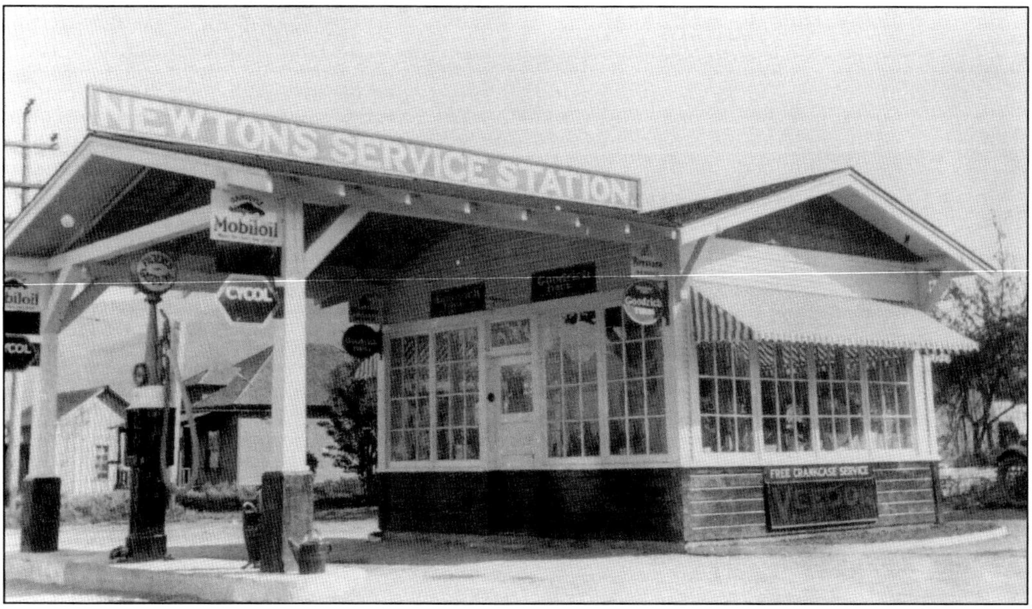

Once there were three service stations along the Pacific Highway in Talent. Newton's Service Station and the Talent Service Station just a block away were both built at the same time in 1924. The third station was located north of town near Suncrest Road. Newton's is now the Downtowne Coffee Shop on the corner of Talent Avenue and Wagner Street.

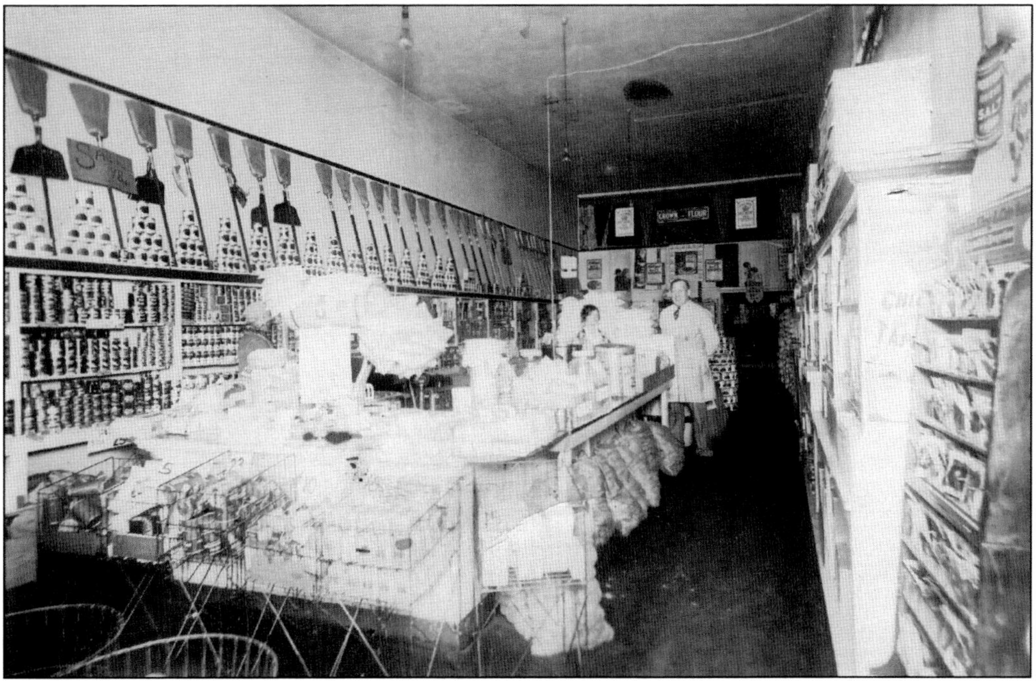

Chester Wolters followed his father's tradition and opened a store in Talent in 1917. This interior snapshot shows Chester Wolters inside what may have been his Talent store in the concrete Wolters building on Wagner Street. (Jerald McGrew.)

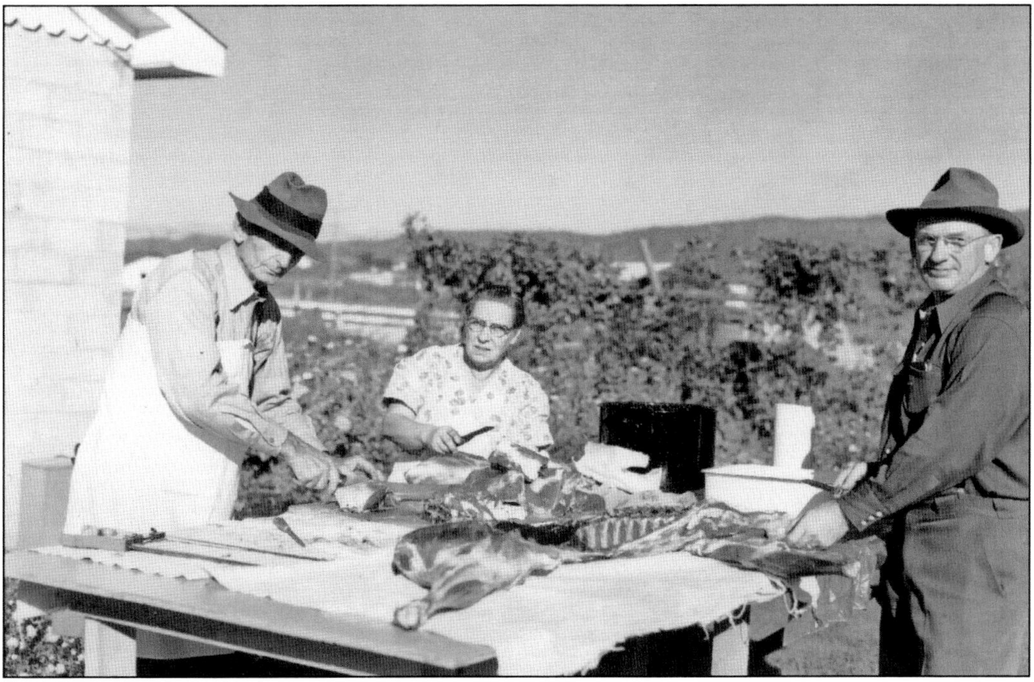

William Childers once ran a butcher shop in town. He and his wife, Ethel, were using their skills to cut enough meat to fill their freezer. An unidentified neighbor (on the right) was given a knife to help with the outdoor butchering session.

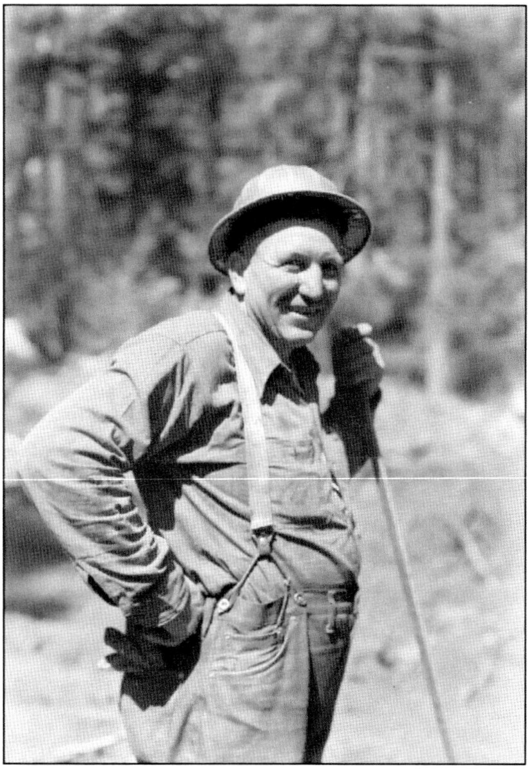

Nearly all available workers in Talent at one time or another were employed by Charles Skeeters at the Skeeters Logging Company of Talent. Skeeters kept the men in line with salty language and worked them long hours logging with heavy modern equipment. (Harry Gleim.)

At Rikter Mountain near Tiller, Oregon, in 1951 were Les Mason, Frank Nettic (with the ax), Will Gleim, George Gleim, and Bill Wolgamott, who all worked for Skeeters Logging Company. Frank lived in a little shack on the Skeeters commercial property. He used his ax to cut the limbs off the downed trees. The Mack truck had a McVay "A" frame log loader mounted on the back. (Harry Gleim)

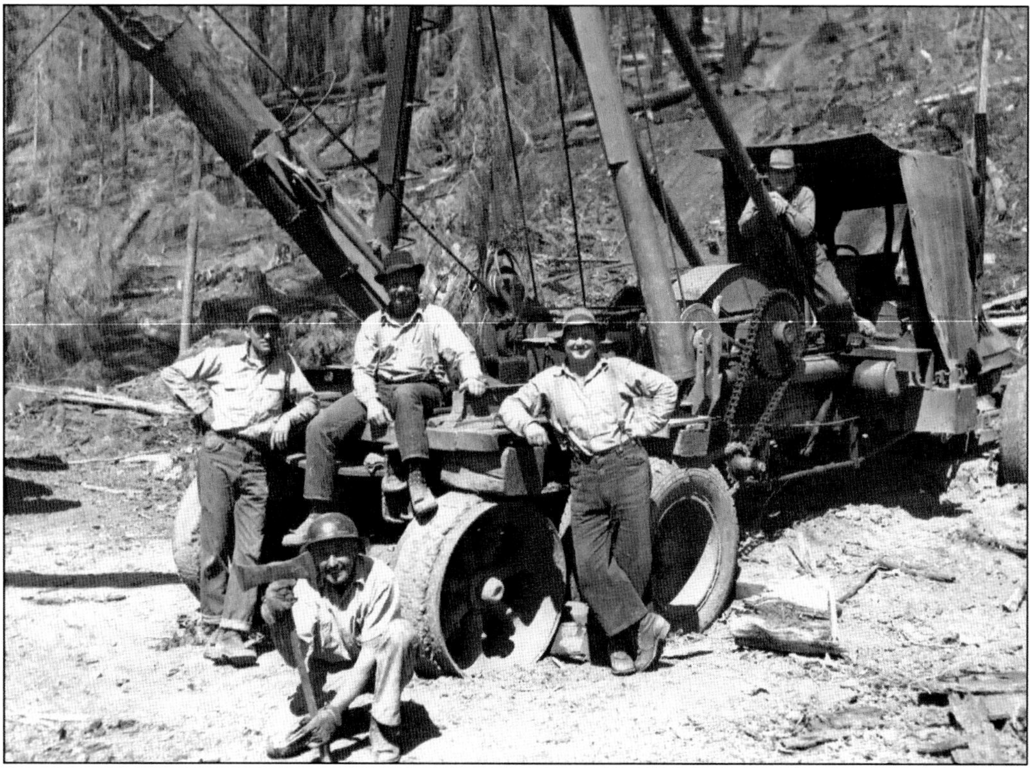

Six

Orchards

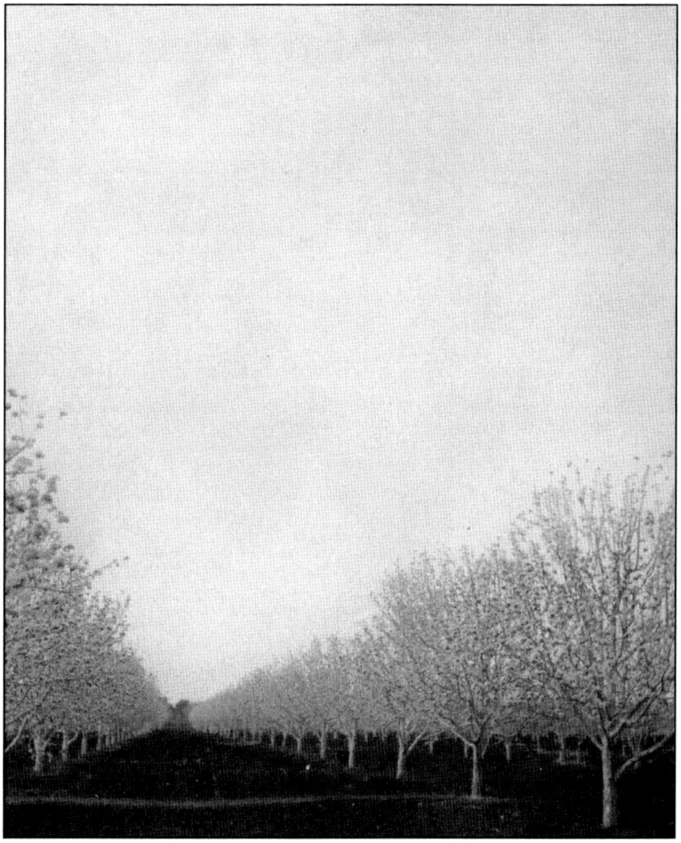

Although the beauty of the orchards made it look easy to just pick the ripest fruit and sell it to waiting customers, any orchardist will agree that frost, insects, wind, hail, too much or too little water, inexperienced workers, transportation delays, excessive heat, packing problems, and fickle consumers could mean disaster. The orchards have been such a big part of the Talent experience that they deserve a chapter of their own. (Jerry Kime.)

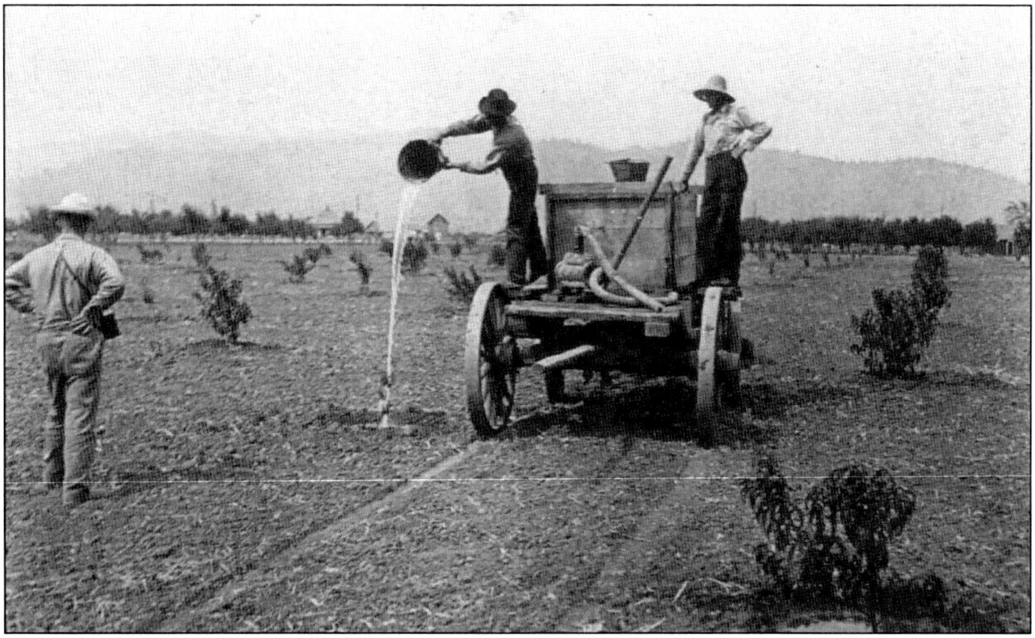

Making a profit took years of work and a number of employees with a number of skills, from hand watering the baby trees to transporting the product to market in carefully packed boxes. When everyone worked together, it worked in a big way. In 1904, Martin Pellett shipped out 13 railroad cars of apples worth $11,000. (Jerry Kime.)

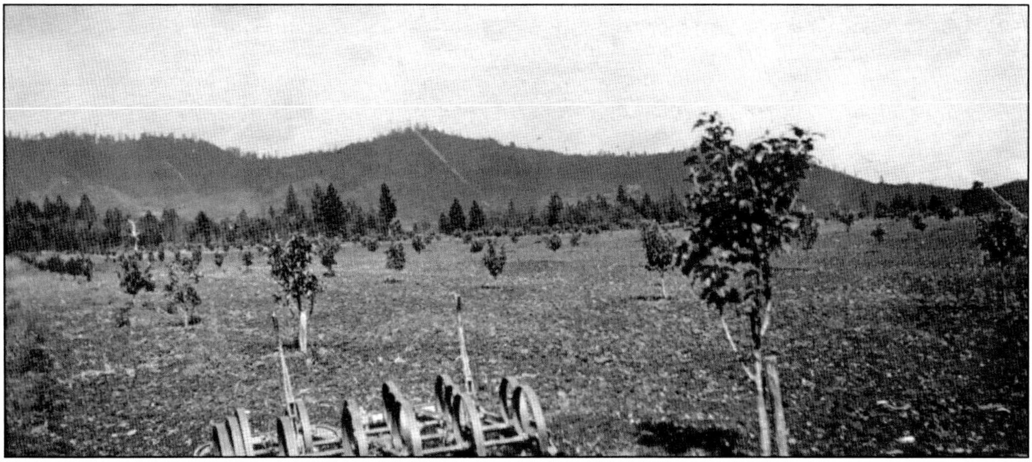

When the Gleims found they needed to cultivate different varieties of fruit, they spaced the trees farther apart than orchardist do now. Not only was it cheaper to buy fewer trees, but mowing and cultivating equipment could easily maneuver between the trees. (Harry Gleim.)

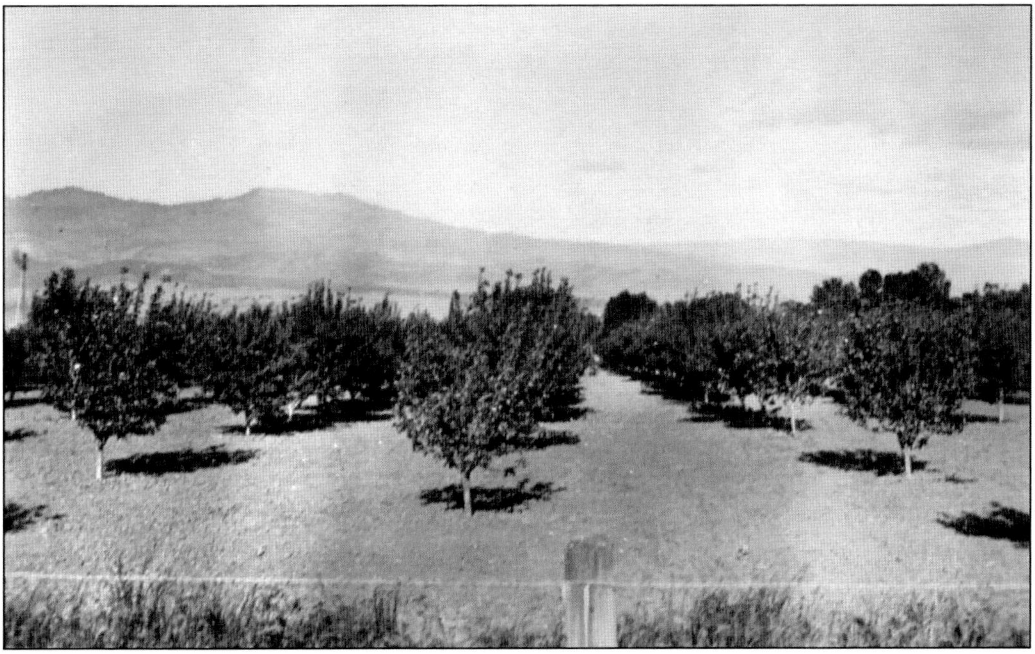

Consumer preferences sometimes changed faster than the fruit could grow. New orchards had to be planted at the right time to take advantage of consumer choice and avoid any insect infestations that could wipe out entire crops. This photograph was of a Gleim orchard on the east side of Wagner Creek Road in the 1920s. (Harry Gleim.)

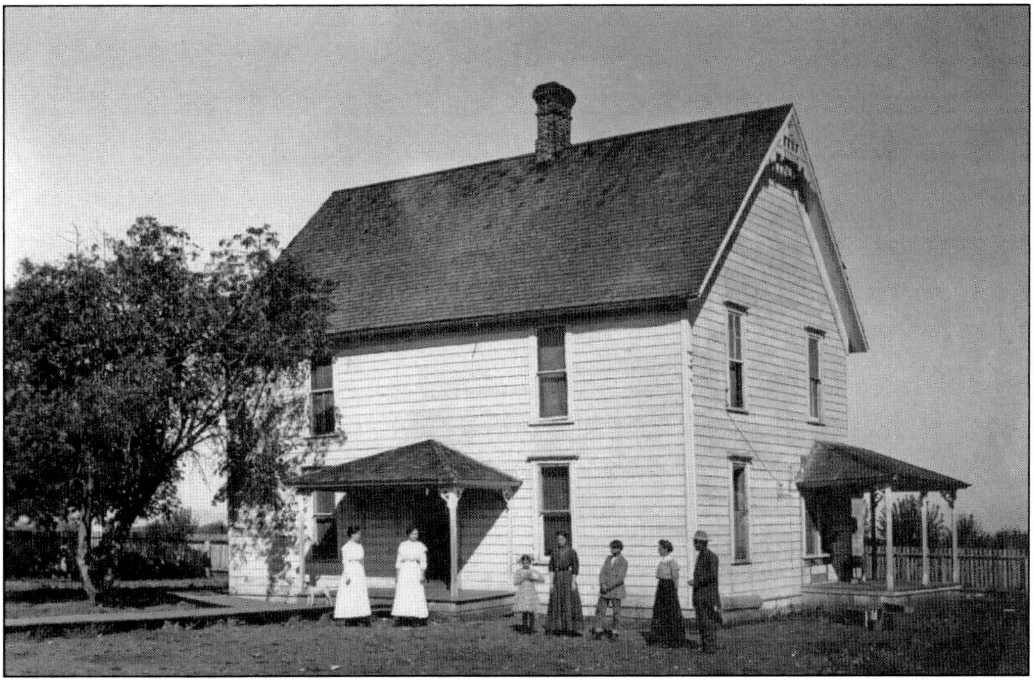

This house was referred to as an orchard house and was probably used for boarding seasonal workers during the harvest. Women were hired to feed the boarders and to keep the place neat and tidy. (Jerry Kime.)

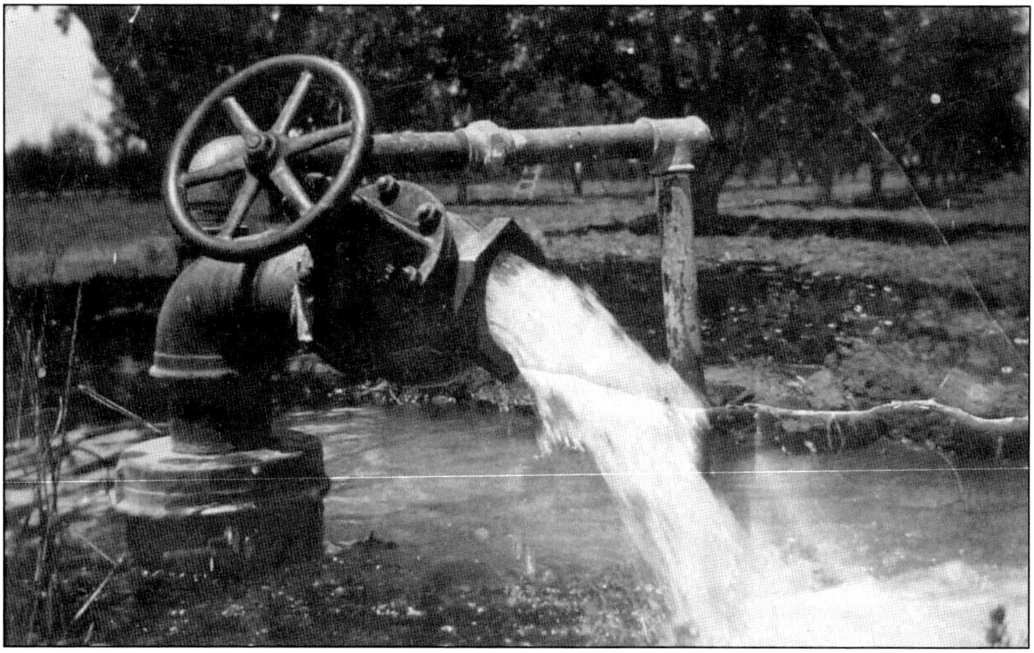

A unique attraction off of Colver Road was a free-flowing spring everyone called the artesian well. The site was a popular spot for bringing a tin cup and getting a refreshing drink. Before 1910, Martin Pellett tried to expand the capacity and the profitability of the well with a bundle of dynamite. Unfortunately, the well collapsed on itself and never again resurfaced. (Jerry Kime.)

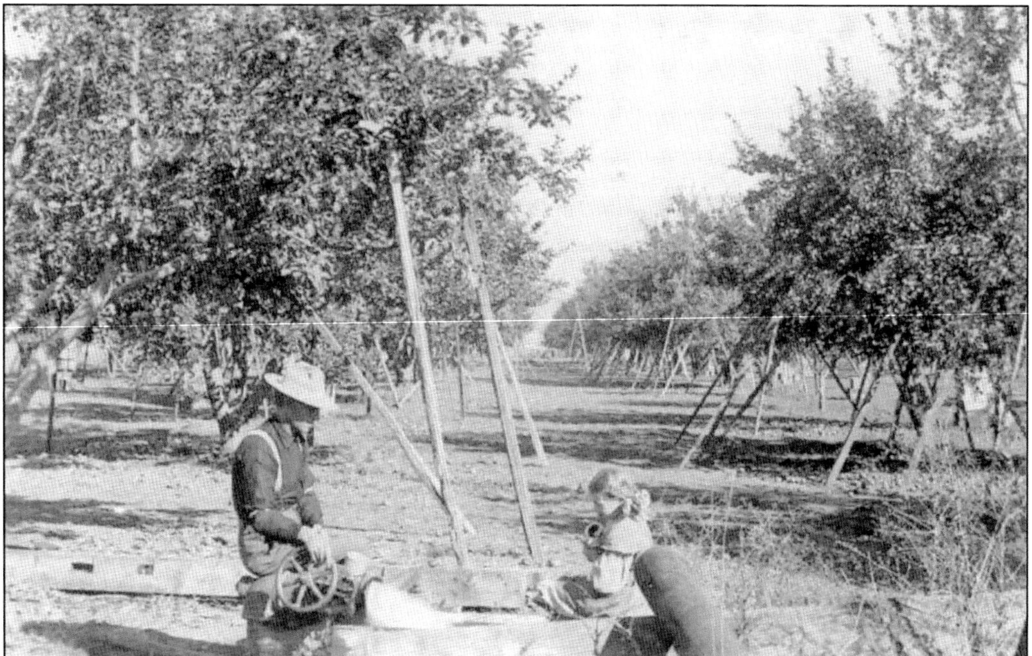

William Joseph Quackenbush and his daughter, Mae, were photographed at the artesian well about 1910. Mae lived to be 107 years old; perhaps the clean and clear waters contributed to her long life. (Jerry Kime.)

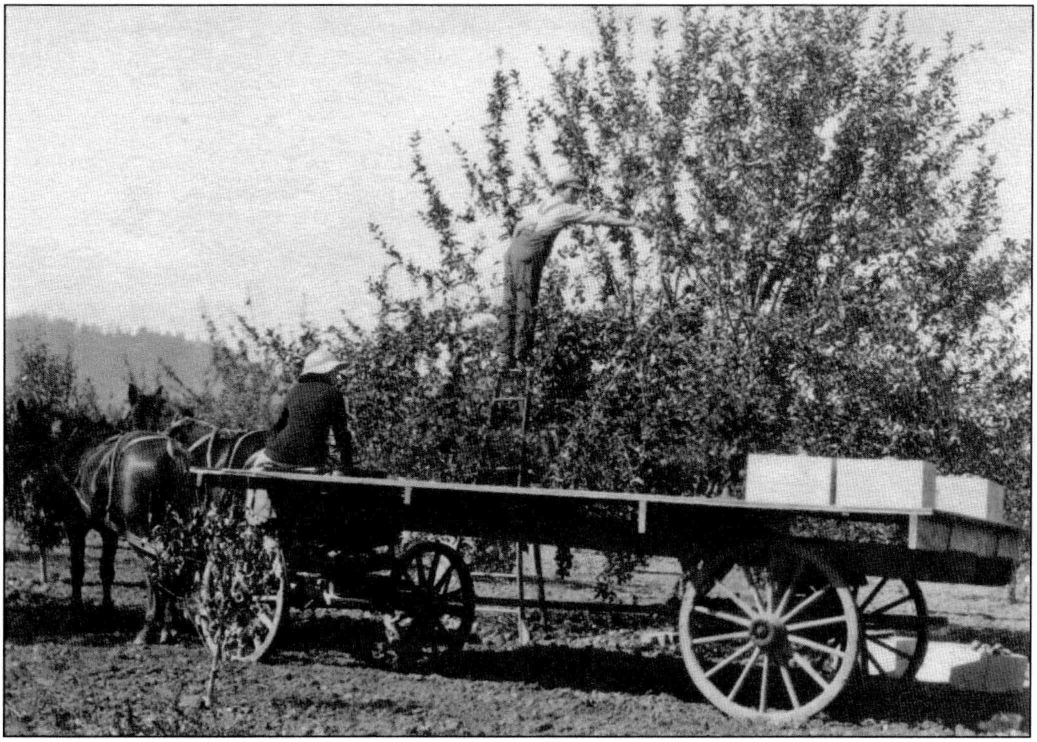

Cora and Will Gleim took advantage of the mature trees that had been growing on the property when Will's parents bought the place in 1908. Apples grew better in cooler climates, so when an insect infestation affected the output, the Gleims had to cultivate more land and plant pears to replace the apples. Some of their orchards are now part of Bear Creek Orchards owned by the Harry and David Company. (Harry Gleim.)

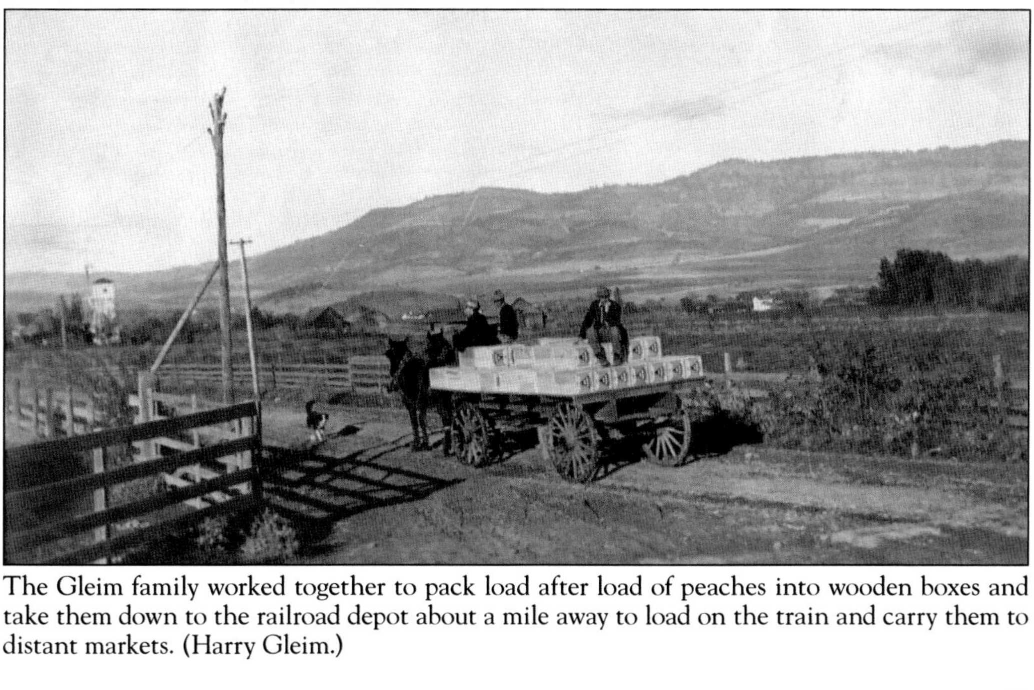

The Gleim family worked together to pack load after load of peaches into wooden boxes and take them down to the railroad depot about a mile away to load on the train and carry them to distant markets. (Harry Gleim.)

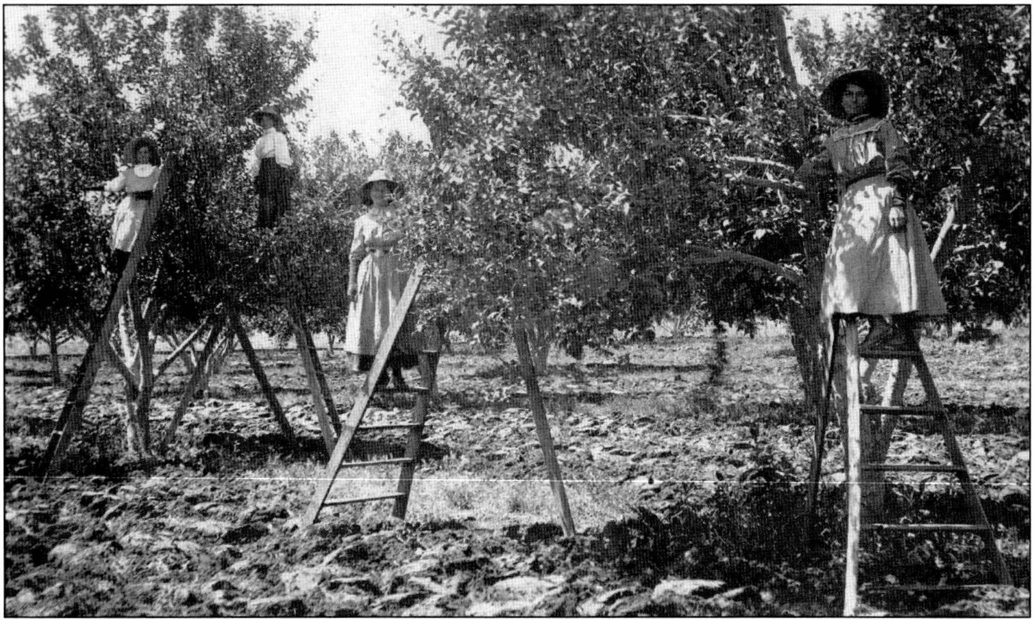

When Eli K. Anderson had a 65-acre commercial orchard in Talent, he employed women and men and paid them the same wages if they did equal work. Later orchardists were not so enlightened. The women they employed received less pay than the men. (Jerry Kime.)

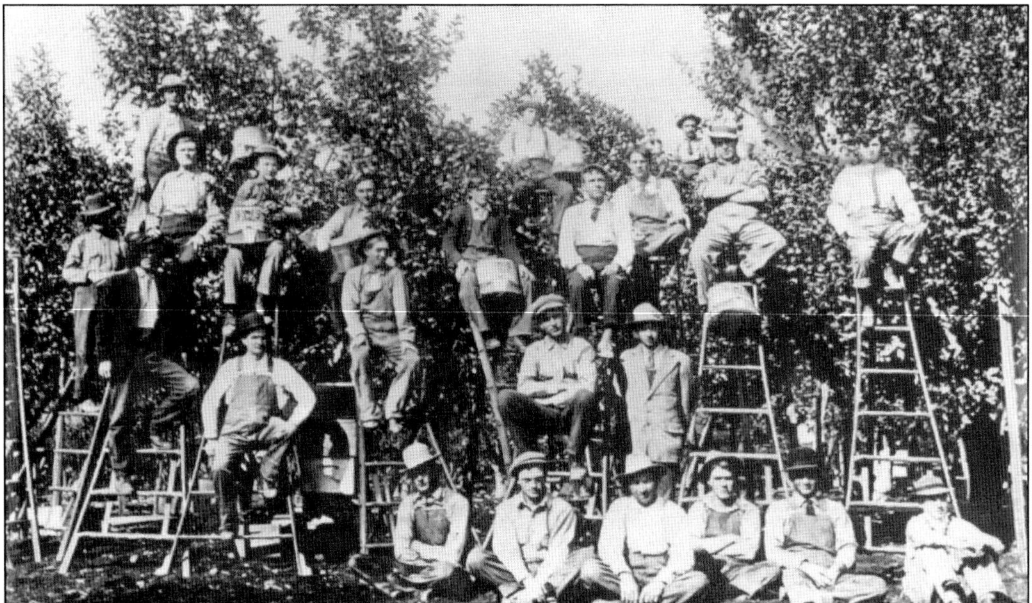

The equipment used in the orchards now has changed, but each pear, apple, or peach is still picked by hand. Most hands-on orchard work is now done by people from Mexico or other Central American countries. Packinghouses and big operations like Harry and David in Medford employ a number of local people to produce, pick, prune, package, and sell their products. (Jerry Kime.)

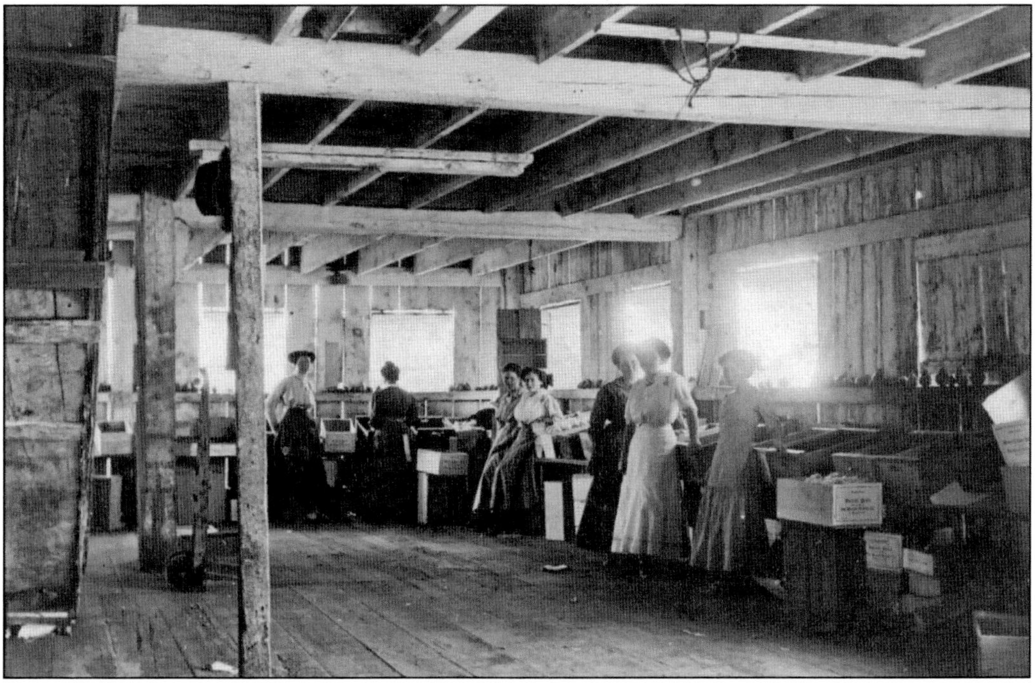

During the boom times, the orchards employed nearly everyone, male and female. Women seemed to have a knack for packing fruit, which made all the difference in getting loyal customers to buy from year to year. This image was taken about 1910 in the Burrell Orchard packinghouse. (Jerry Kime.)

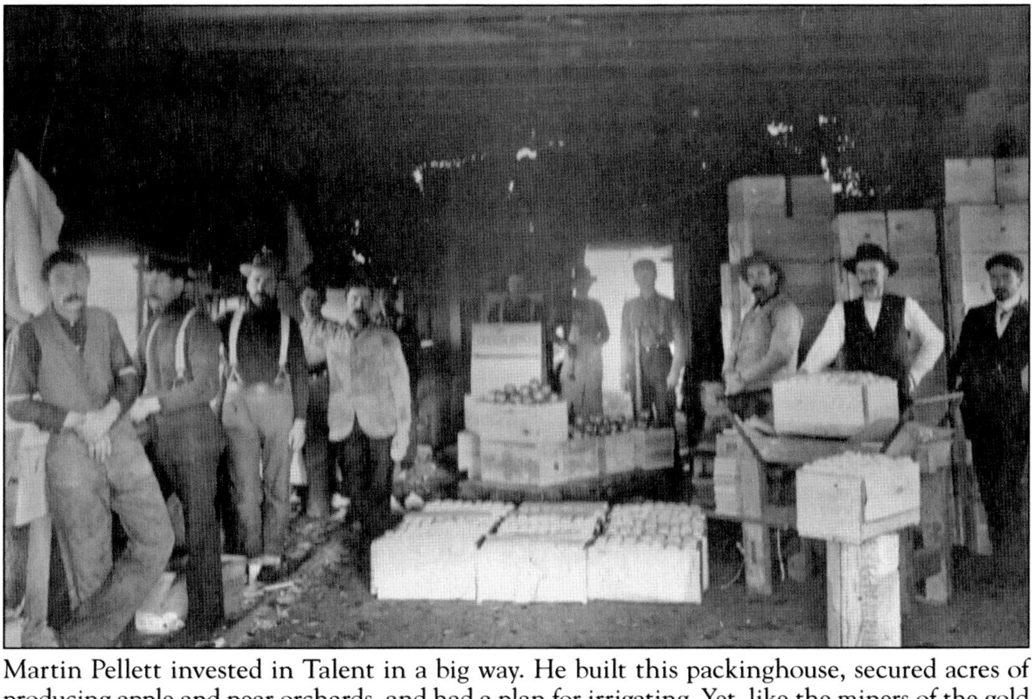

Martin Pellett invested in Talent in a big way. He built this packinghouse, secured acres of producing apple and pear orchards, and had a plan for irrigating. Yet, like the miners of the gold rush days, he moved on when the profit wasn't enough for him. (Jerald McGrew.)

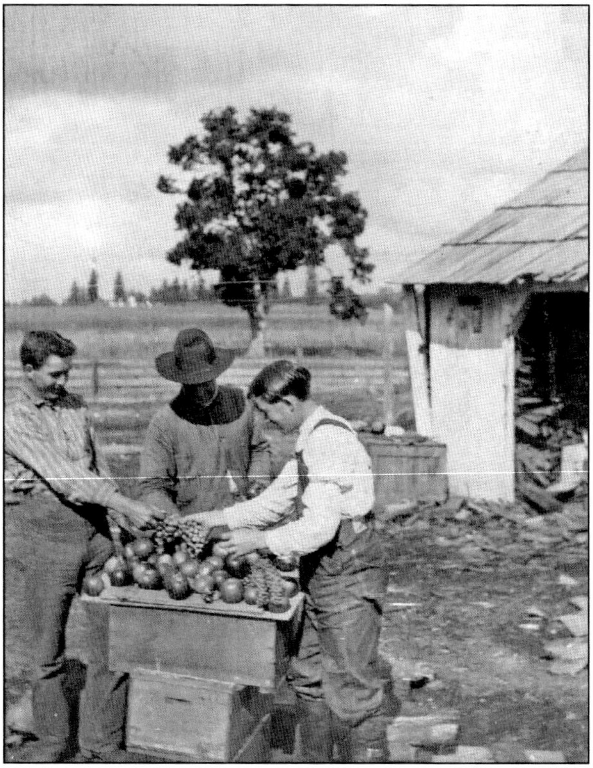

The Quackenbush family of Talent took time out from the harvest to show off their bounty. Before pears became the most popular fruit grown in the Rogue Valley, Talent was best known for its peaches and apples. Plums, grapes, and nuts were also part of the early harvests. (Jerry Kime.)

Three unidentified beauties posed in the Gleim peach orchard on Wagner Creek Road around 1920. Many years before, the first peach to ripen on Wagner Creek was grown on the David Stearns property. Although he was offered $5 for the peach, he decided instead to divide it seven ways and feed a slice to each member of his family. (Harry Gleim.)

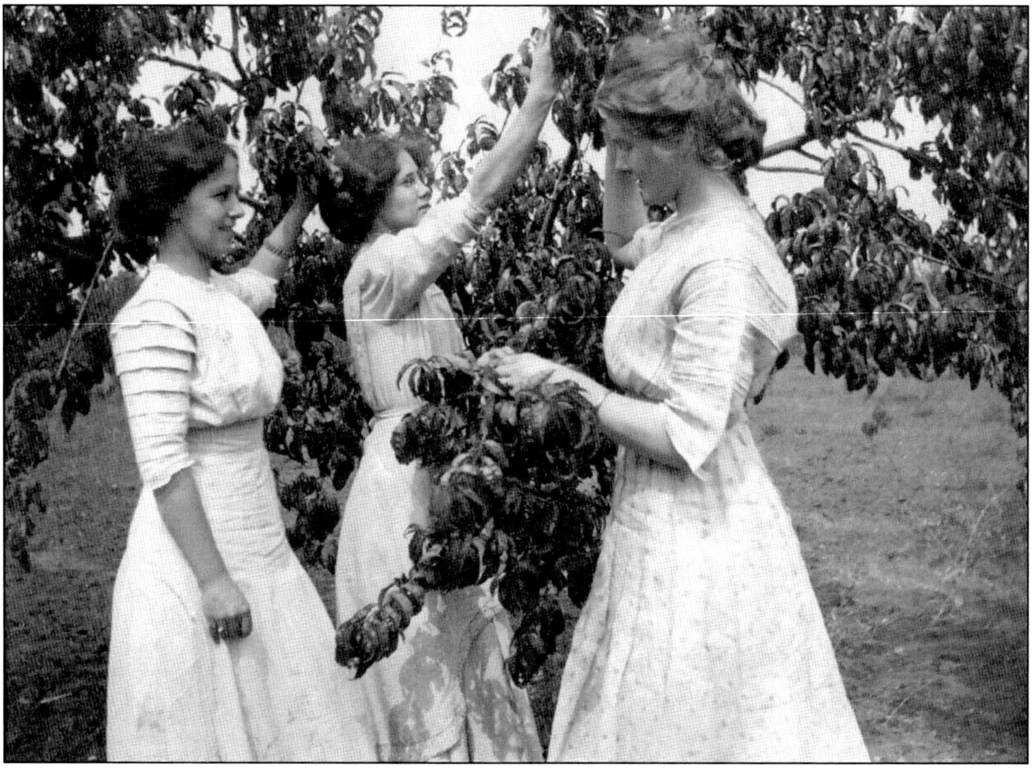

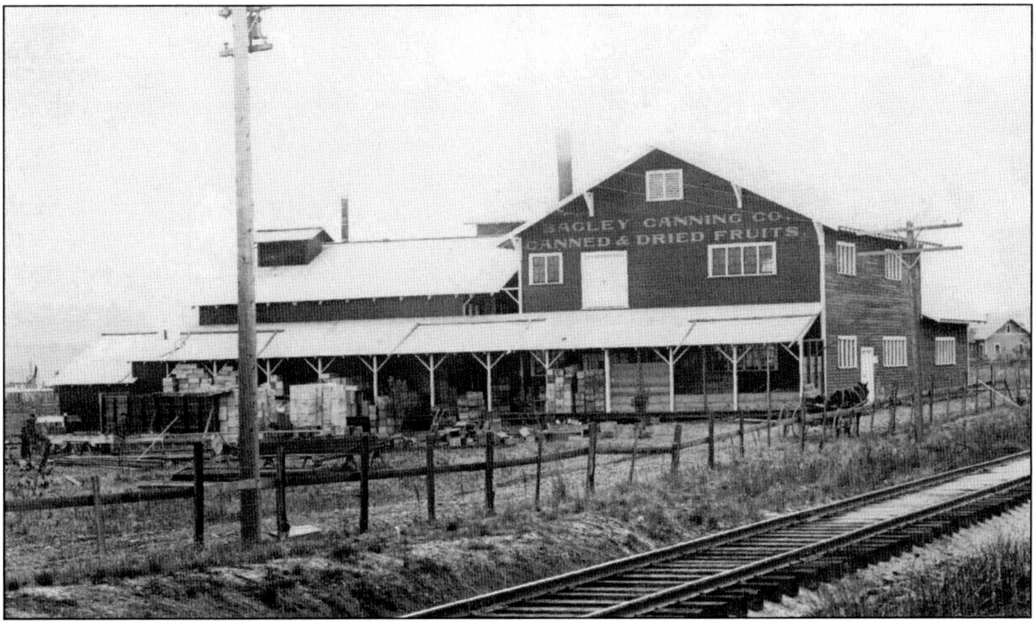

The Bagley Cannery was built in 1912 along the railroad tracks with ready access to markets. At first, it canned mostly tomatoes and peaches but later expanded to other products. The cannery was a source of pride and income for Talent residents until the business was sold and moved to Ashland in 1924. (Jerry Kime.)

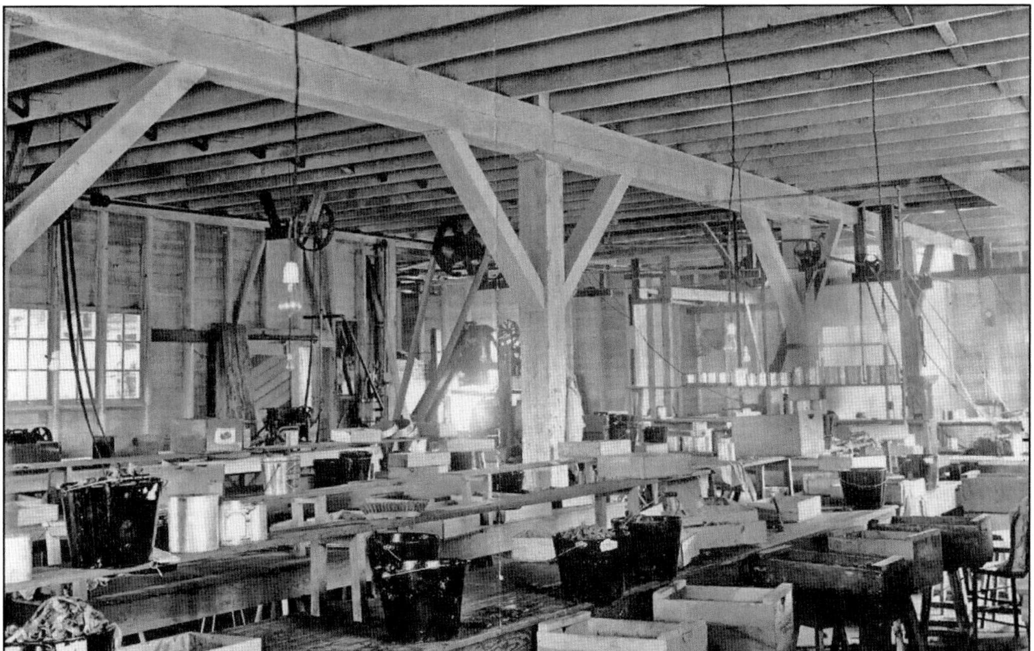

When he came from Minnesota, Dr. William Bagley was very well received in this neighborhood. As one neighbor observed, Bagley enjoyed knocking around among the cows and chickens and working in the garden and fields. The interior of the Bagley Cannery in 1914 was organized and clean enough to satisfy a doctor. (Jim Maize.)

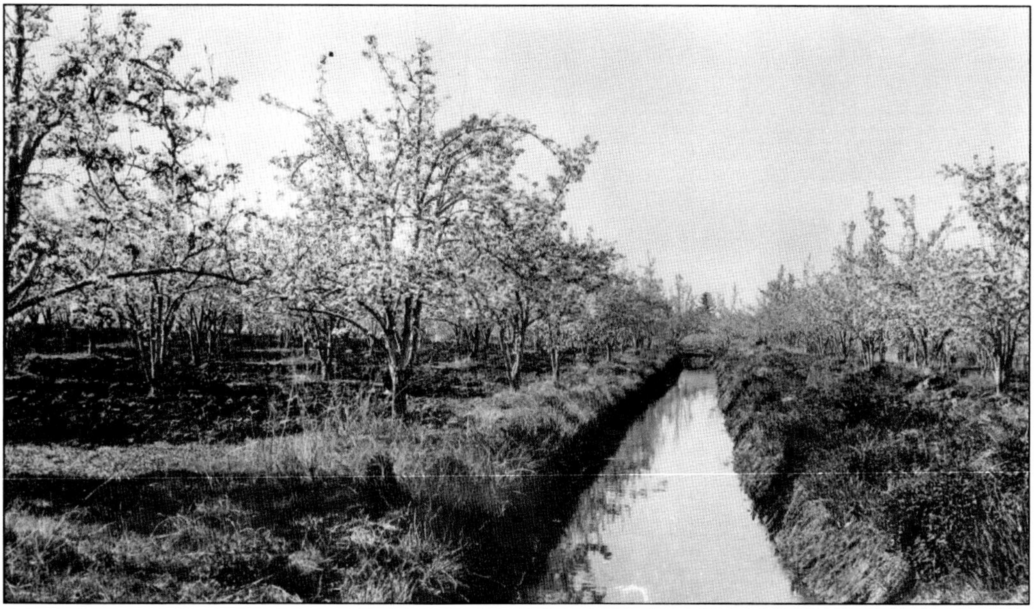

Talent Irrigation District supplies the valley with much needed water for orchards and farmland. Jacob Wagner of Talent is credited with being the first to divert water from Wagner Creek to irrigate local farmland. Disputes over water rights between the miners and the farmers were common during the early settlement days. Water rights continue to be highly prized.

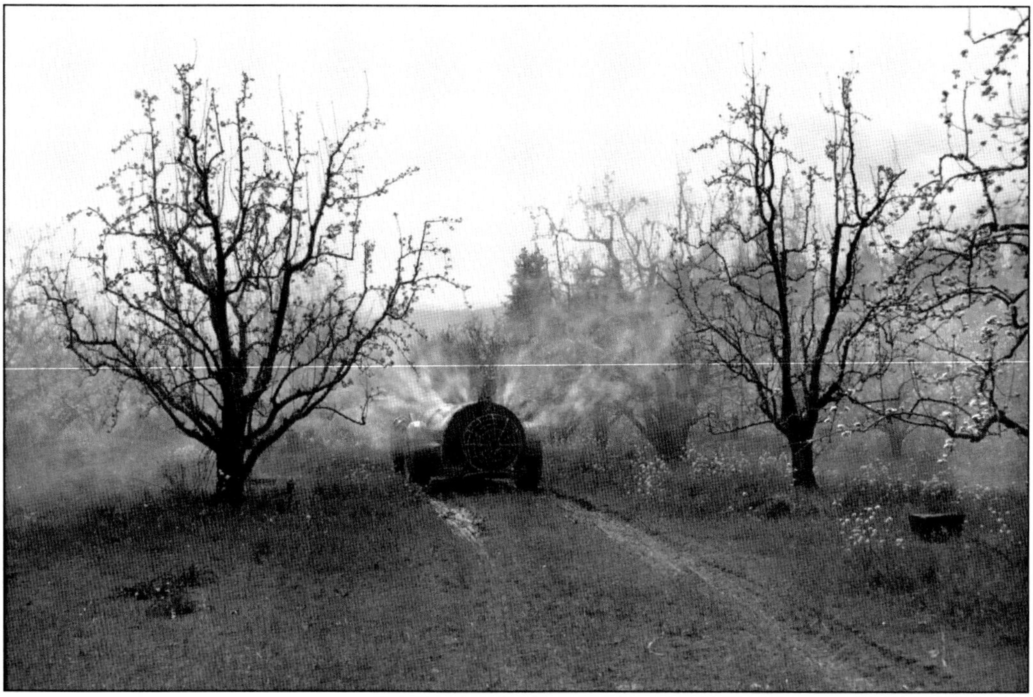

Farmers were known to use soapy water to wash their trees before chemical sprays came into use. Chemicals were developed when bugs and disease started to threaten the crops, and the use of them was not as controversial as it is now. There are some experimental organic orchards today, but the majority of the growers prefer to use chemical sprays to protect their crops. (Ken Kantor.)

Seven

SCHOOLS

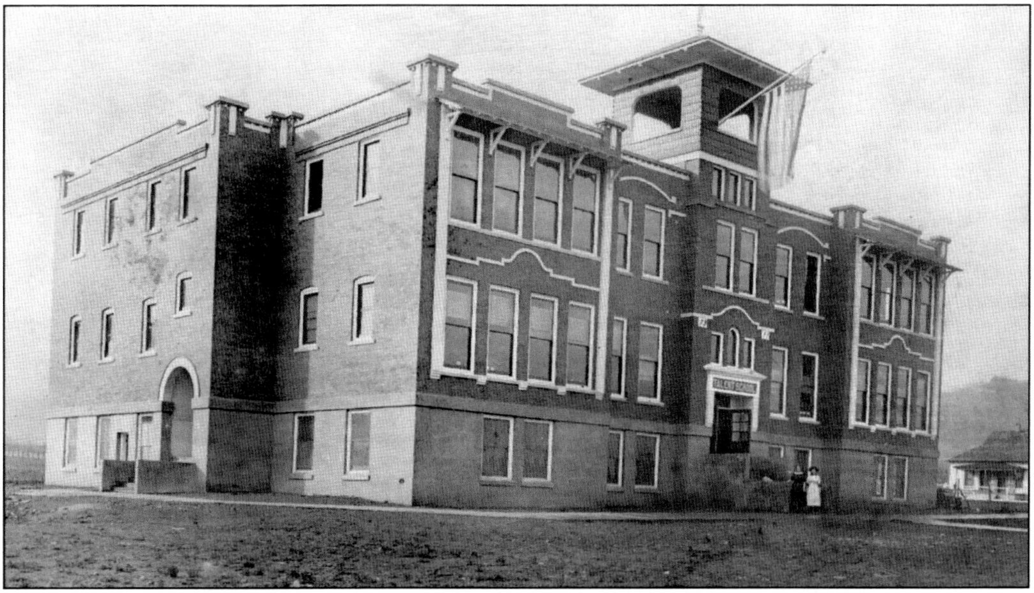

The Talent brick schoolhouse, costing $32,000 and built in 1911, was an impressive landmark for over 60 years. Many memories were made in the school, which are captured in some of the following photographs. A 1914 newspaper article about the school bragged about the building and the management, including manual training teacher Miss Groshong, "who demonstrates the fact that a woman can saw a board, square, and drive a nail without swearing."

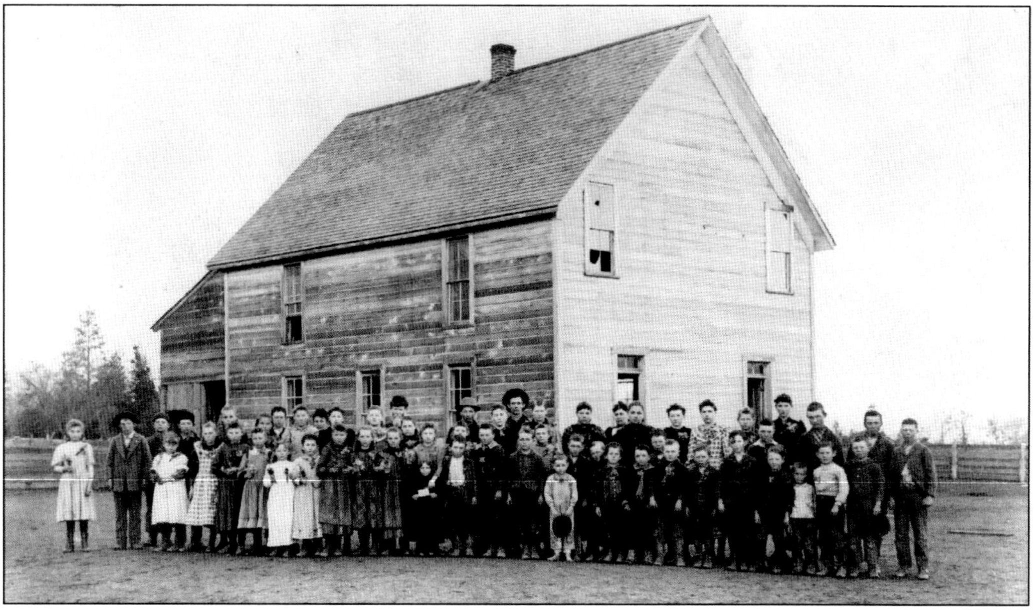

This 1880s photograph may be the schoolhouse referred to as the North Wagner Creek school, located near what is now Beeson Lane. The Beeson family deeded the land to School District 22 in 1879. Even though some children came to school barefooted, not a single blade of grass survived the trampling feet on the play yard. (SOHS No. 8589.)

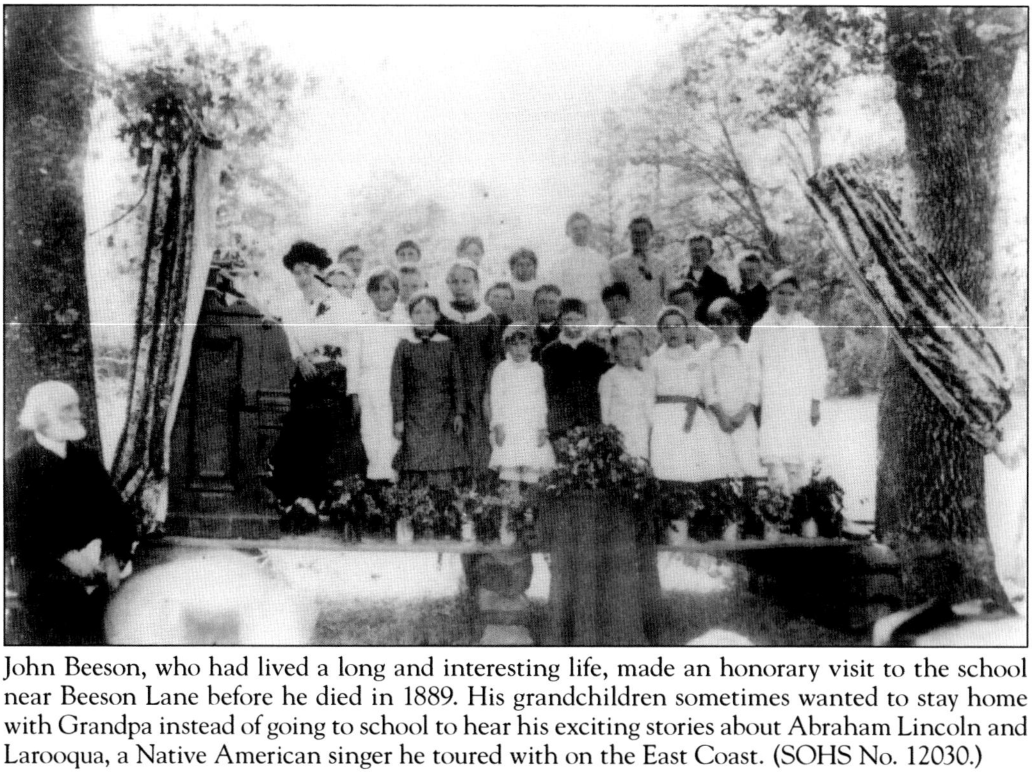

John Beeson, who had lived a long and interesting life, made an honorary visit to the school near Beeson Lane before he died in 1889. His grandchildren sometimes wanted to stay home with Grandpa instead of going to school to hear his exciting stories about Abraham Lincoln and Larooqua, a Native American singer he toured with on the East Coast. (SOHS No. 12030.)

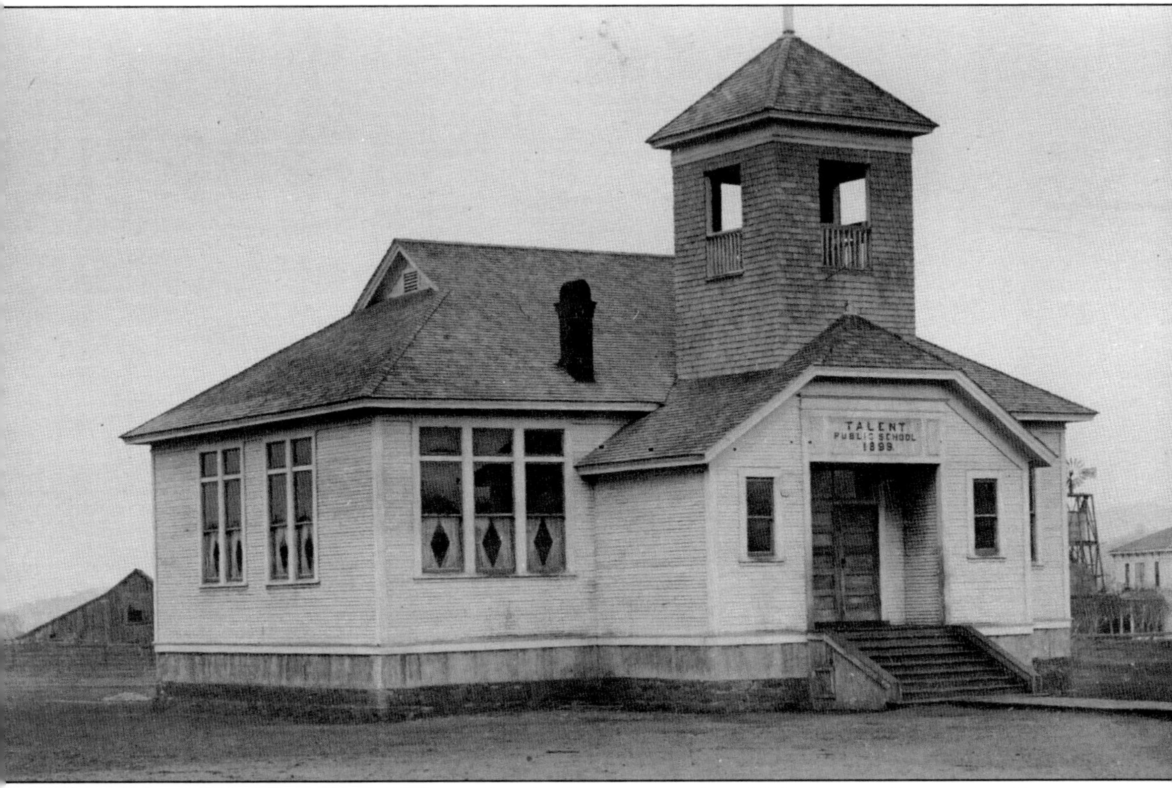

What is now the Community Hall was first built as a school in 1899. It served as a schoolhouse until 1911, when population increases called for a new school building. It has been used as a dance hall, as a library, for church meetings, for city business, and as a venue for community clubs and events. The school bell still hangs in the tower, but the ringing rope is out of reach.

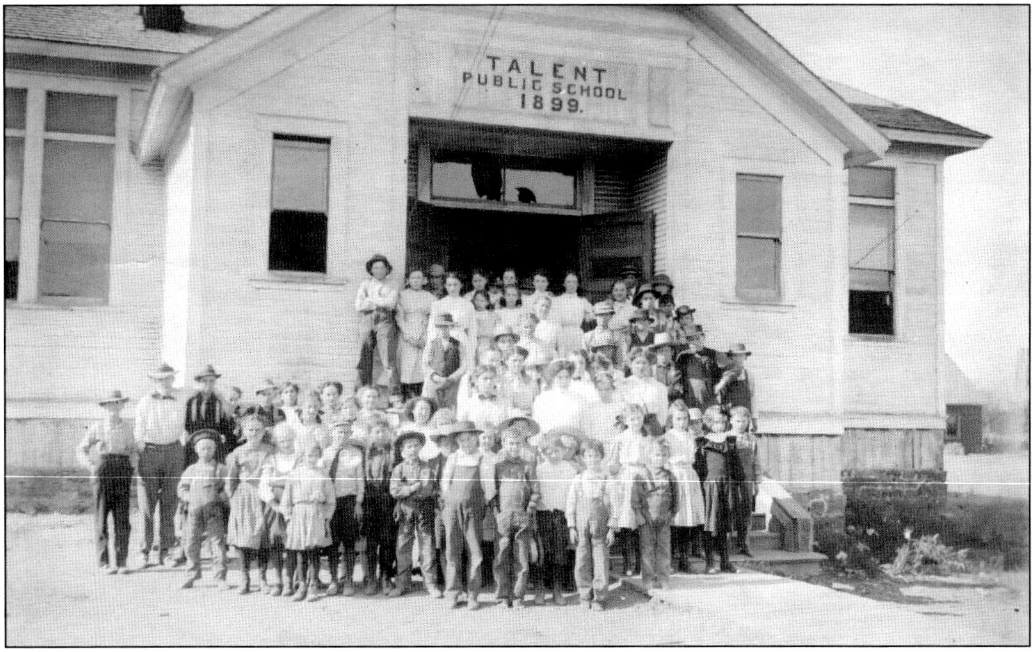

The 1899 school on Main Street was a two-room schoolhouse with the older children on one side and the younger children on the other. Teachers had to practice strict discipline to keep the noise from spilling into the other classroom. It is evident from this photograph that Talent was attracting families with young children and that a new school would soon be needed.

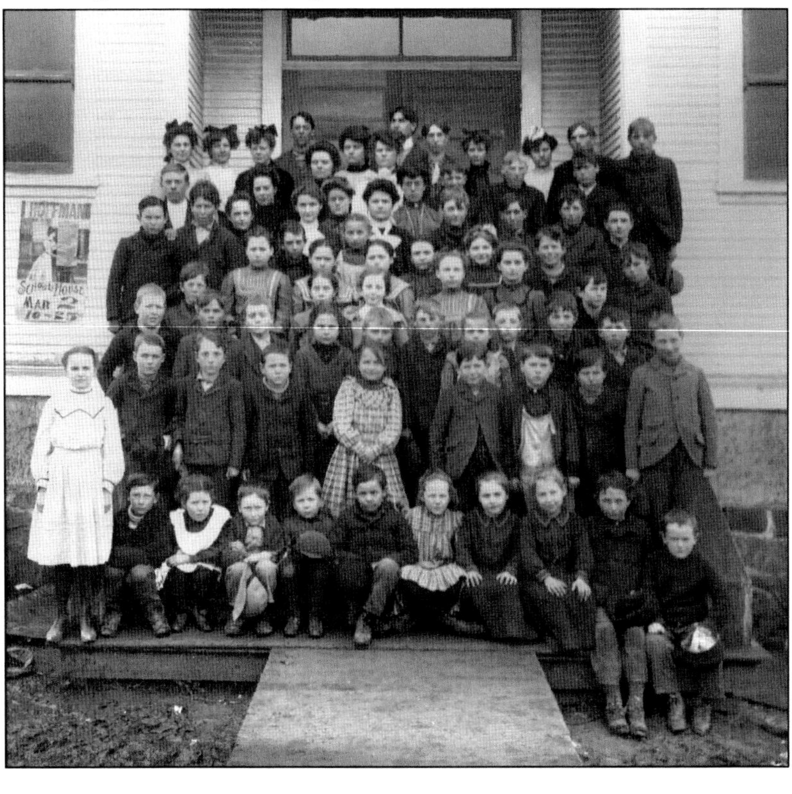

Hairstyles, clothing, shoes, and manners may be different now, but childhood faces are timelessly full of promise and mischief. This photograph was taken in 1903 of the students who attended the Main Street School.

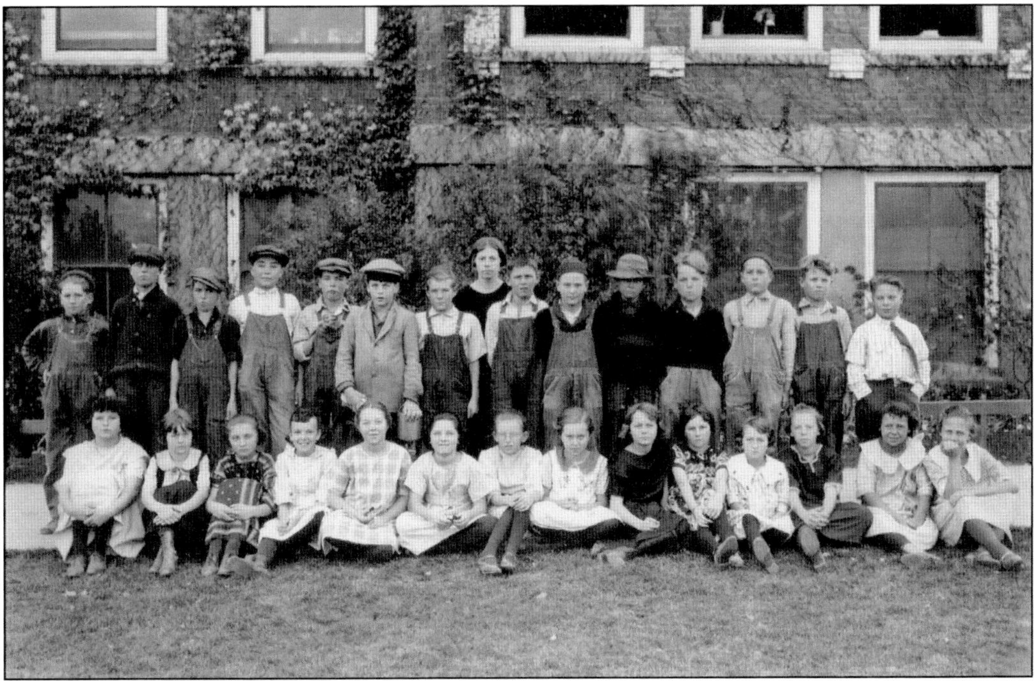

In this 1920s class photograph, the boys look like they can't wait to bust out and play marbles. The girls were challenged to maintain modesty while sitting on the ground in their dresses. This spot on the side of the school was a favorite place for classes to have their pictures taken.

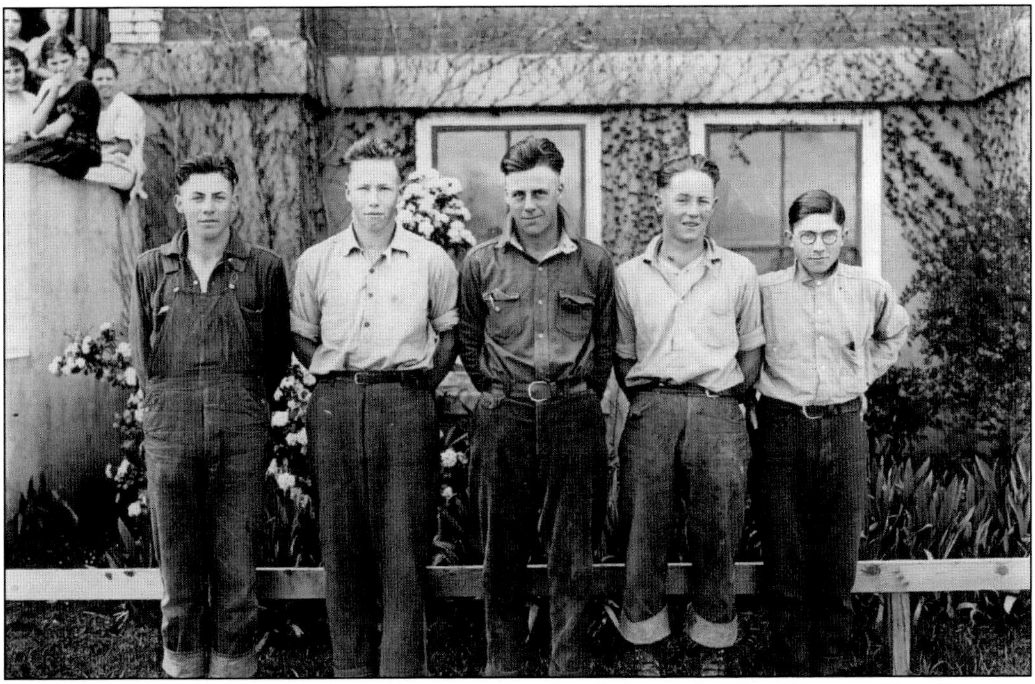

The champions of the 1918 basketball season were, from left to right, Floyd Purves, Glen Seymour, Ellis Beeson, Harold Burnette, and Glasgow Stratton. They were most proud of beating the Applegate team 16-14, Harold Burnette making the winning basket.

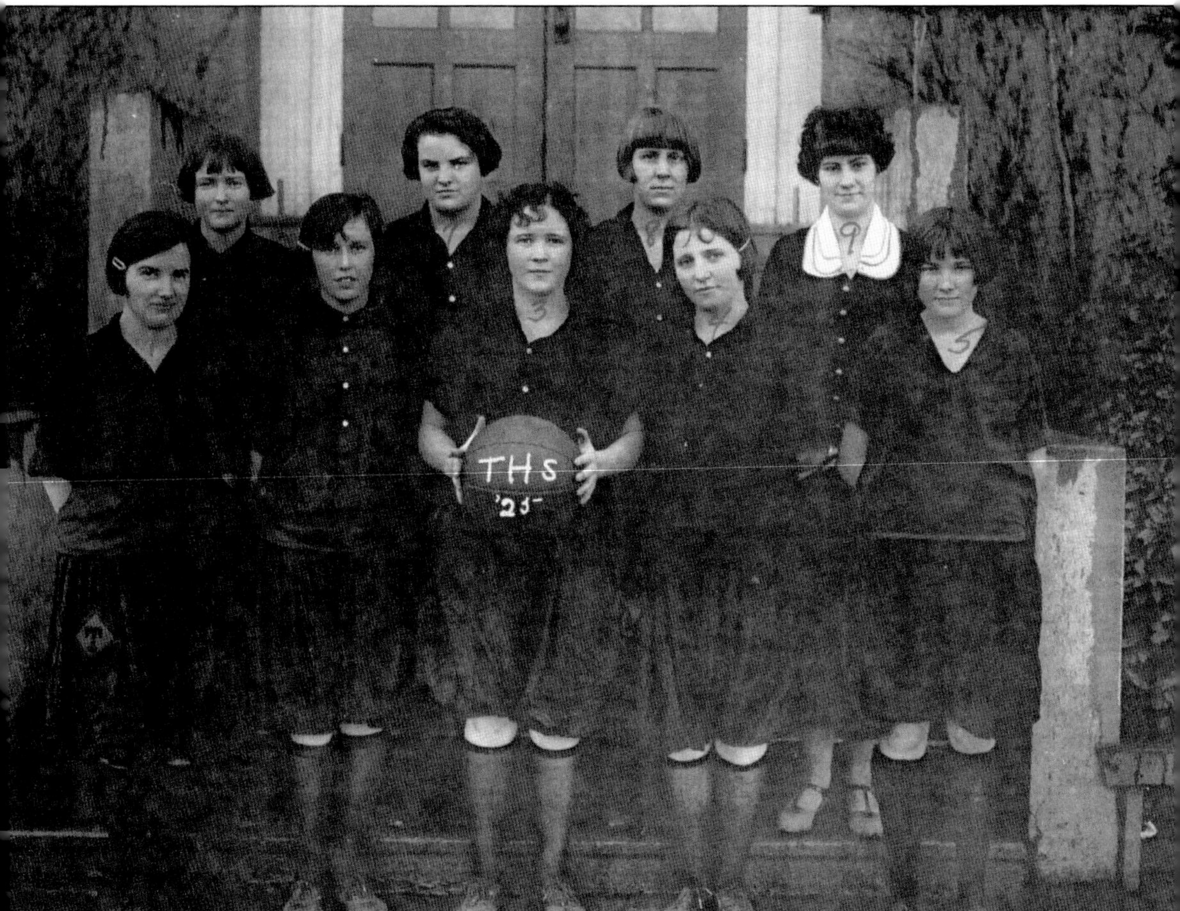

The black uniforms with a bright orange "T" on the bloomers made this 1925 Talent girls basketball team stand out. From left to right are (first row) Betty Brown, Ruth Bowman, Beulah Burnett, Margaret Purves, and Edna Newbry; (second row) Lois Hill, Elsie Newbry, Ada Hackler, and teacher Mrs. Sanders.

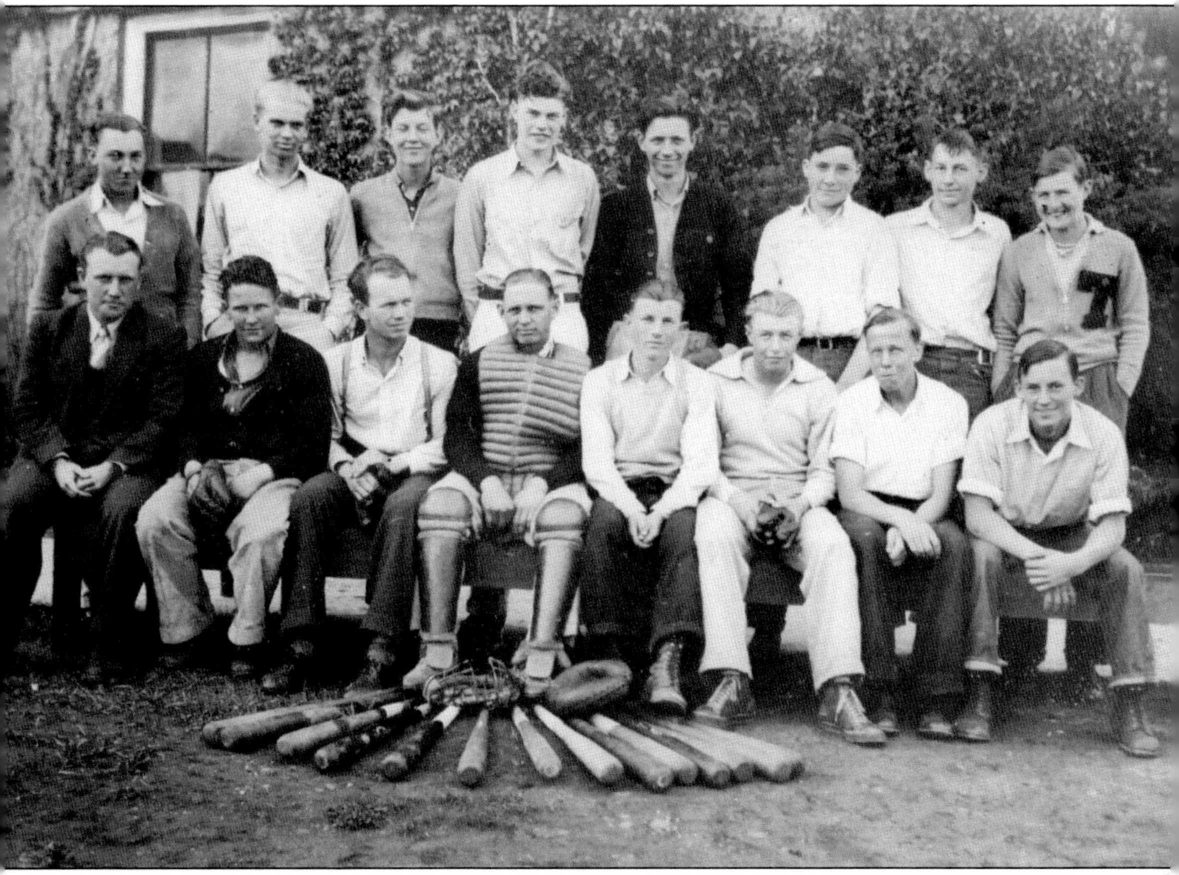

The 1937 Talent High School baseball team had six victories and no defeats that year. They even won all their practice games. John Childers in the catcher's gear successfully caught Don Hungate's fireball pitch and Richard Skeeter's curveball from behind the plate.

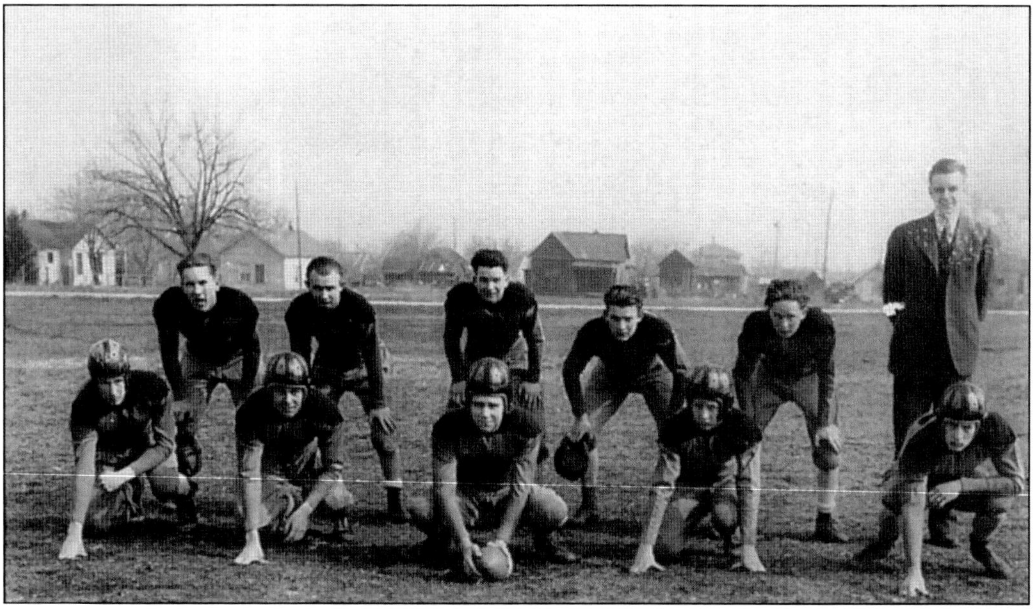

In 1943, the high school football team had only one week's practice when they won their first game against Gold Hill. The highest score they got that year was 40-6 against Prospect. George Gleim was the fullback pictured in the center of the back row. (Harry Gleim.)

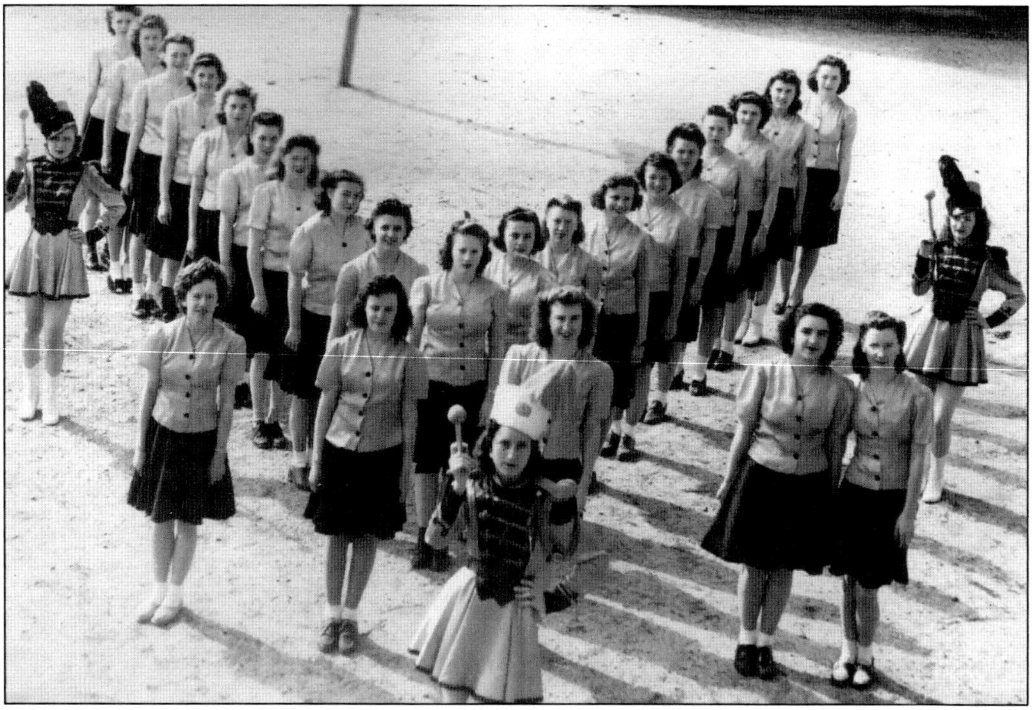

The majorettes who led the 1943 drill team in their orange uniforms were Julia Klimek, Wanda Lou Hill, and Margie Gleim. That year, the Talent Drill Team was trained by the principal, Mr. Roy Parr, when the regular drill team advisor, Mrs. Gooldy, was unable to live in Talent.

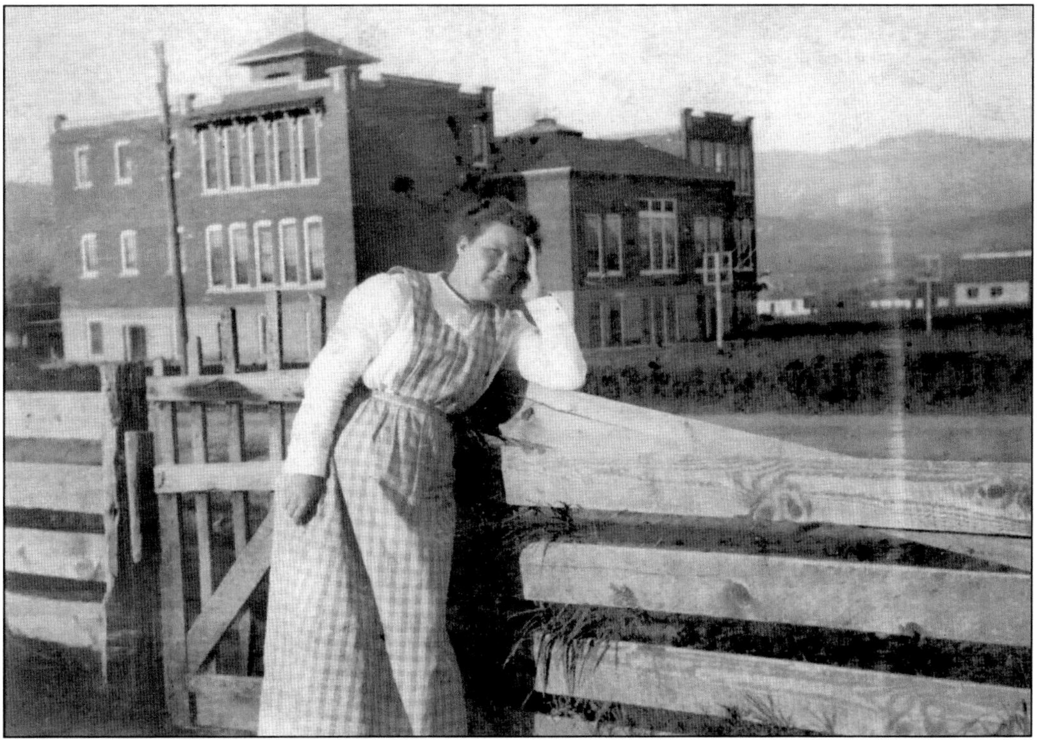

This photograph of Marcia Coleman Robison, who lived on Wagner Street across from the schoolhouse, shows a rare view of the back of the building.

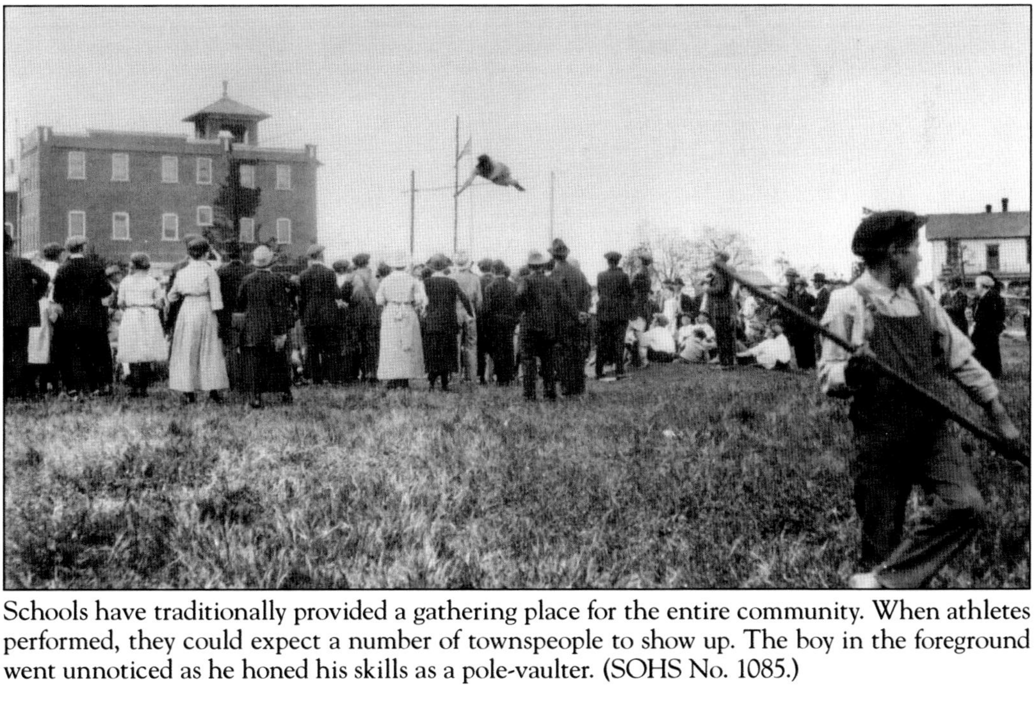

Schools have traditionally provided a gathering place for the entire community. When athletes performed, they could expect a number of townspeople to show up. The boy in the foreground went unnoticed as he honed his skills as a pole-vaulter. (SOHS No. 1085.)

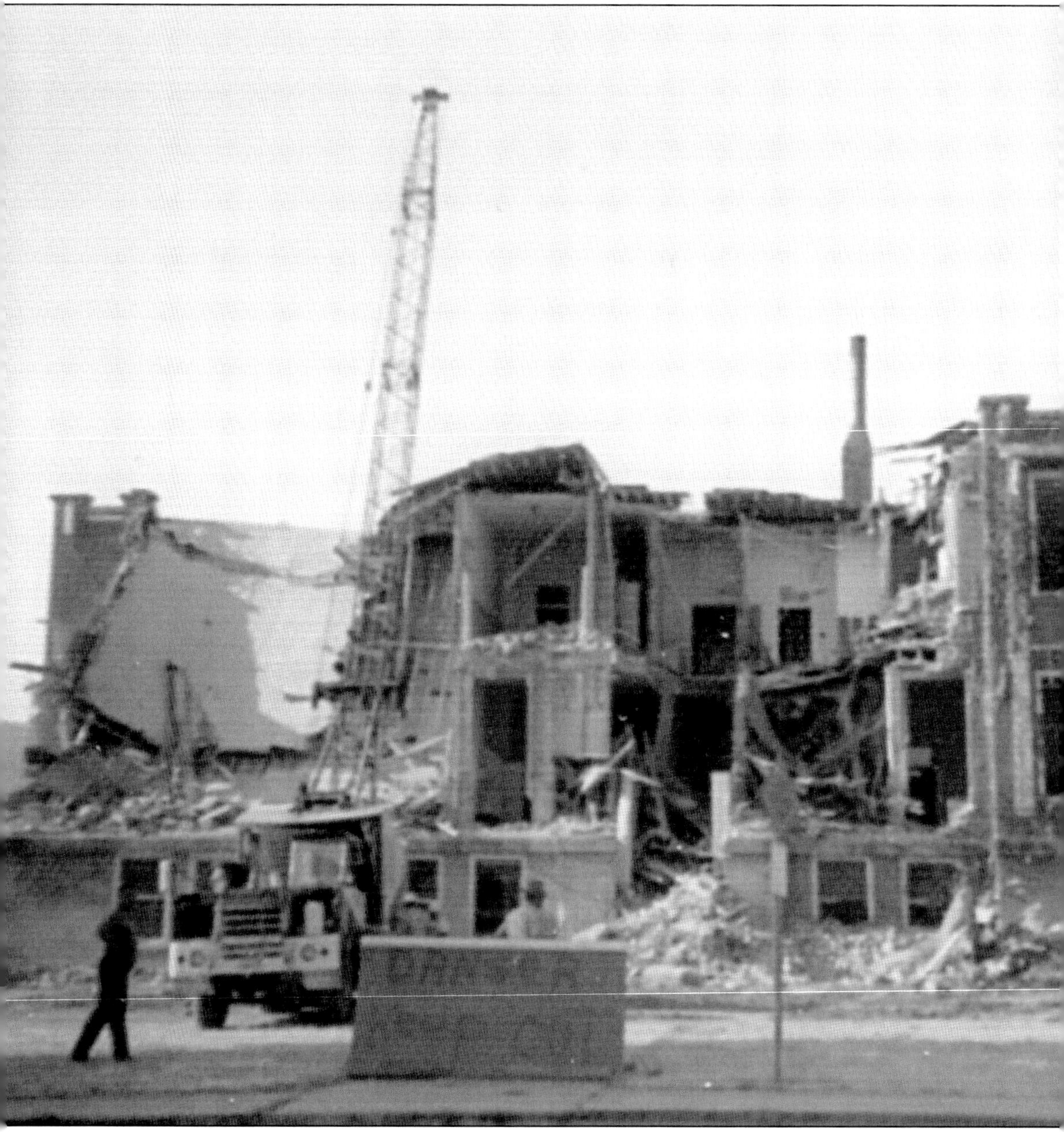

It was bad enough that Talent had to consolidate with Phoenix schools in the 1960s, but insult was added to injury when the old brick schoolhouse was torn down in 1973. Some people saved a few of the bricks just to hang on to the memory of the school they had attended.

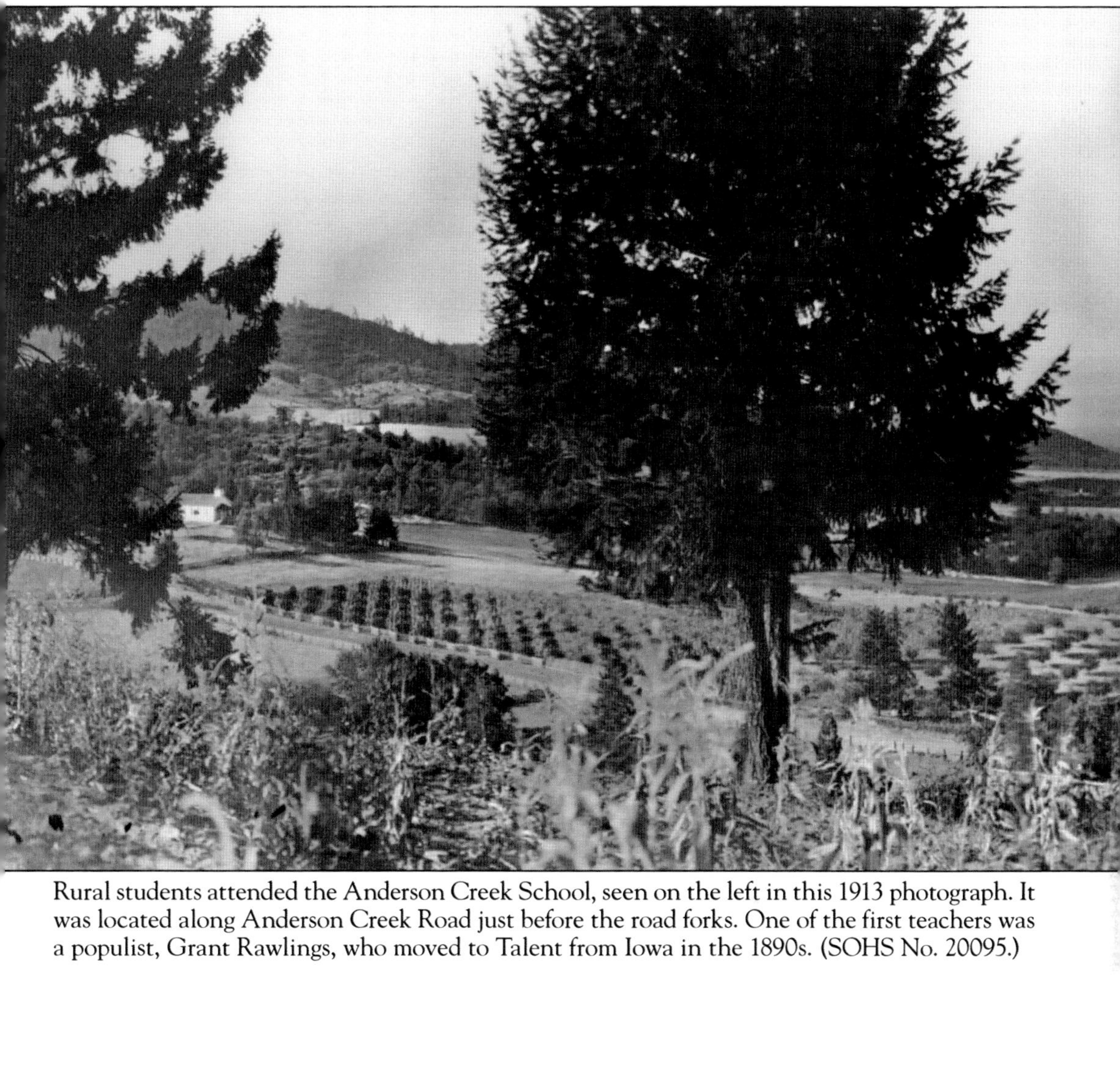

Rural students attended the Anderson Creek School, seen on the left in this 1913 photograph. It was located along Anderson Creek Road just before the road forks. One of the first teachers was a populist, Grant Rawlings, who moved to Talent from Iowa in the 1890s. (SOHS No. 20095.)

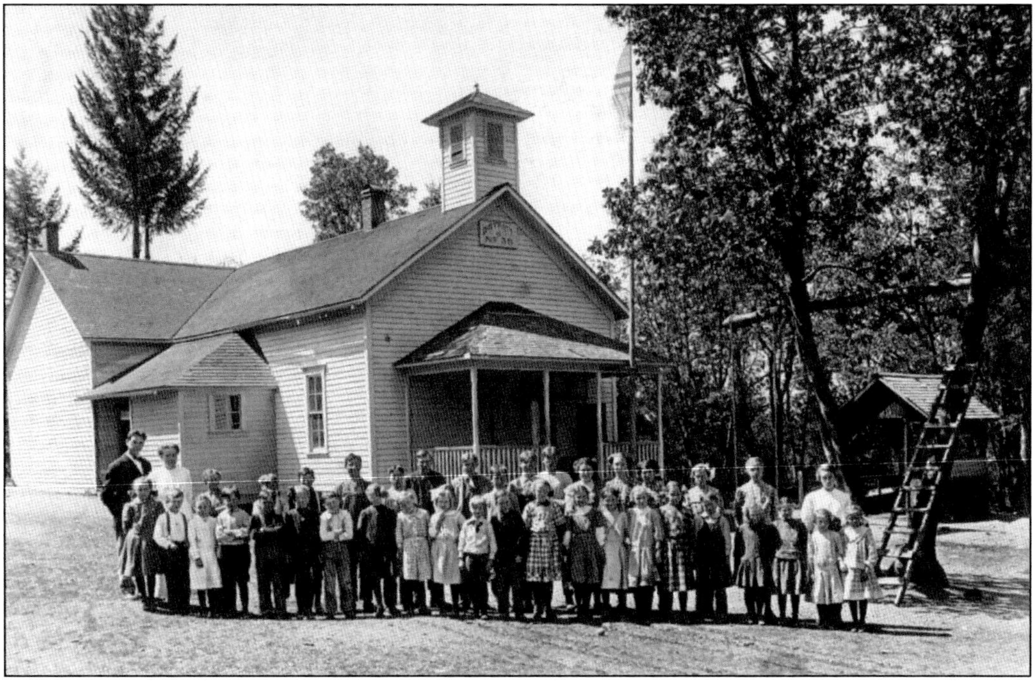

School district No. 56, the Wagner Creek School, was formed in 1872 on the George Lynch property on Wagner Creek Road. Students' surnames in this 1913 image of the school include Beeson, Kerby, Smith, Davis, Purves, Lynch, Abbott, Scott, Briner, and others. Wells and Molly Wheeler were the teachers. (Everett Taylor.)

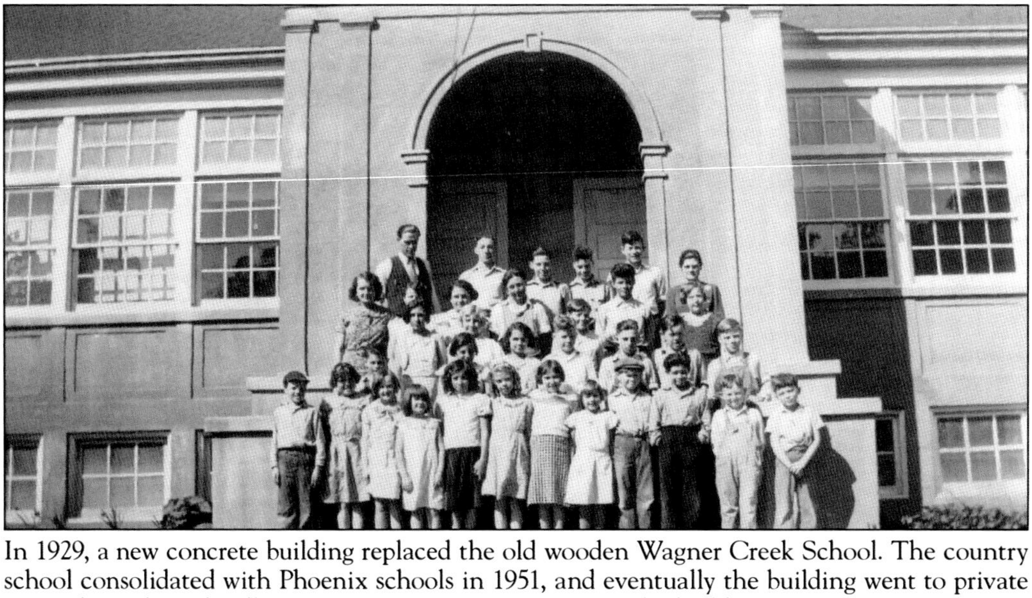

In 1929, a new concrete building replaced the old wooden Wagner Creek School. The country school consolidated with Phoenix schools in 1951, and eventually the building went to private ownership. The Schoolhouse Retreat center now occupies the building.

Children need to play. The primary-room children in 1942 paused and crowded onto the swing set at Wagner Creek School just long enough for the photograph to be taken. They immediately resumed swinging after the click of the camera.

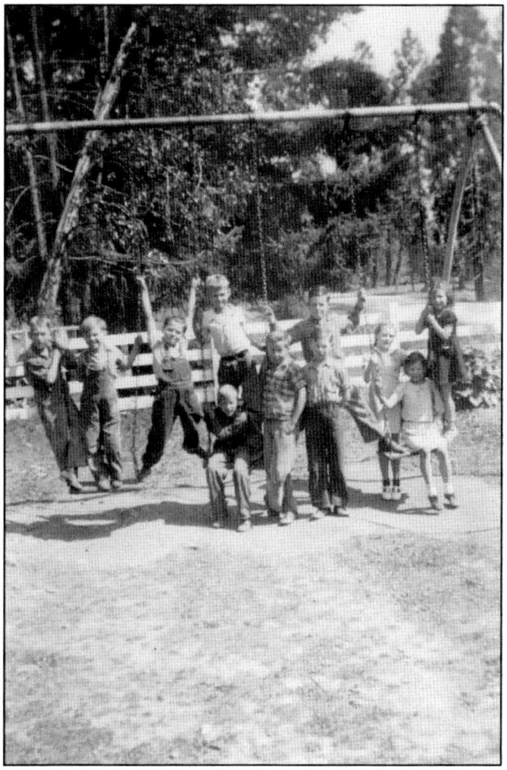

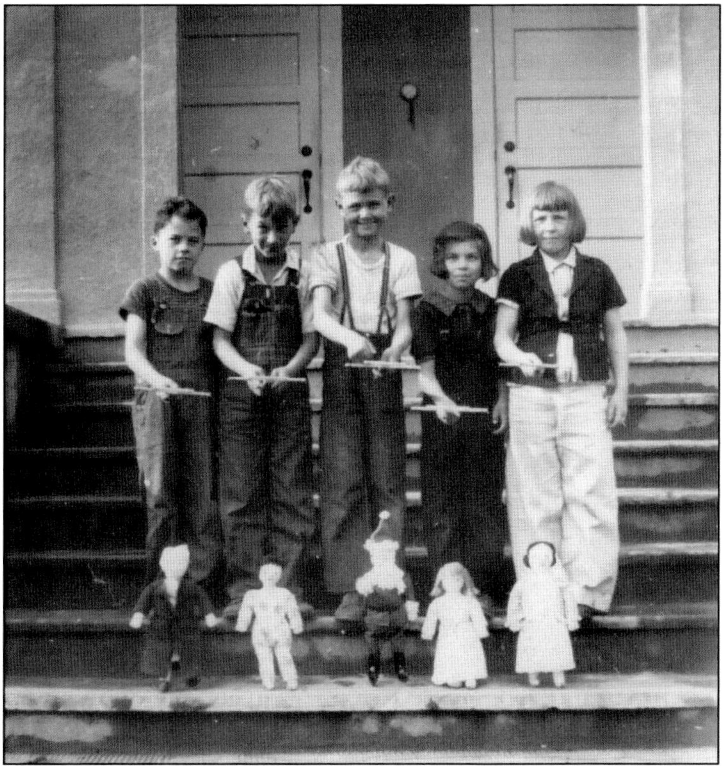

The same primary-room children showed off their puppetry skills with handmade puppets that had been used in the 1941 Christmas show. They probably still remembered their lines from the puppet show by the end of school year, when this picture was taken.

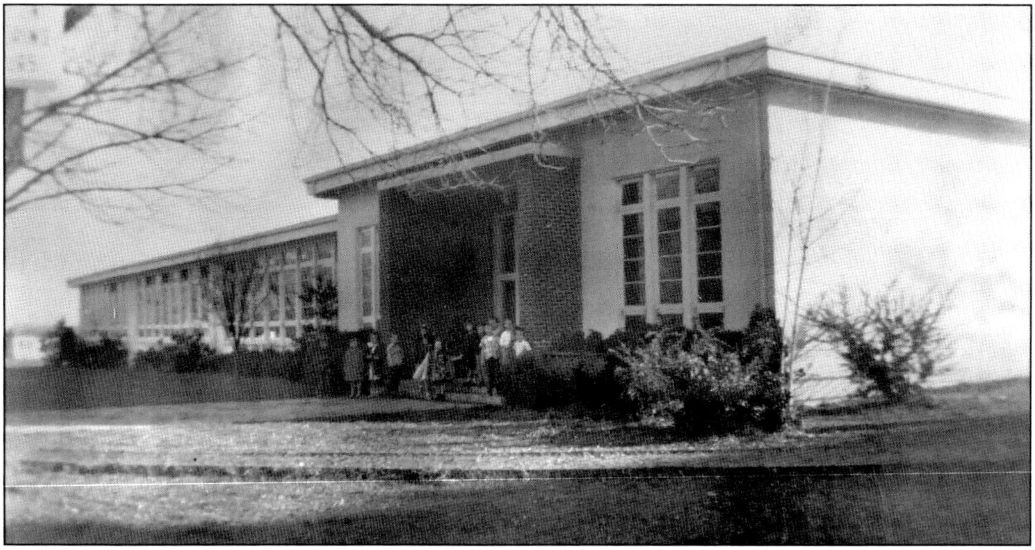

Talent Grade School for grades 1–3 was built in 1949 and is still in use today as part of the Talent Elementary School. When this photograph was taken, the big brick school still stood next door, where grades 4–8, the cafeteria, library, and principal's office were located.

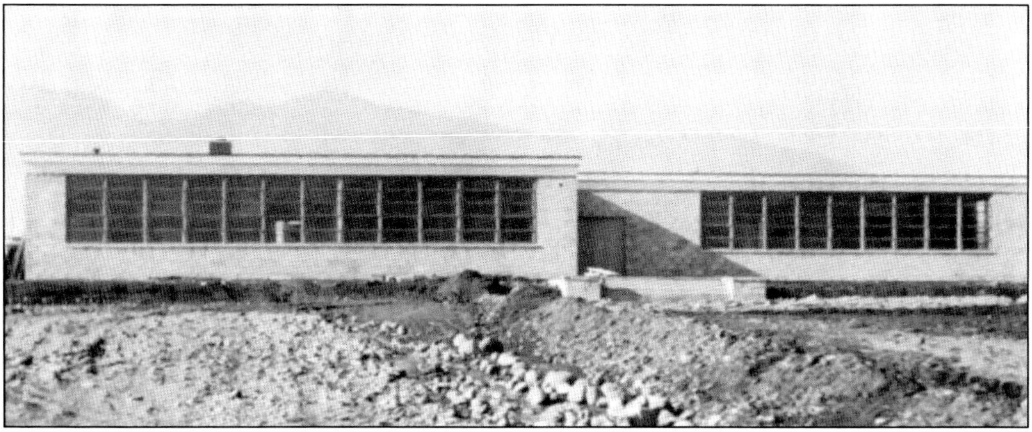

By 1955, the new Talent High School was ready for occupancy. The yearbook for that year had a front-page story with photographs taken during the construction of the building, from excavation to placing the windows. The building boasted spacious halls, generous locker space, and even a teachers' lounge. The interior was painted flamingo, canyon rose, chartreuse, and melody green.

Eight

Recreation

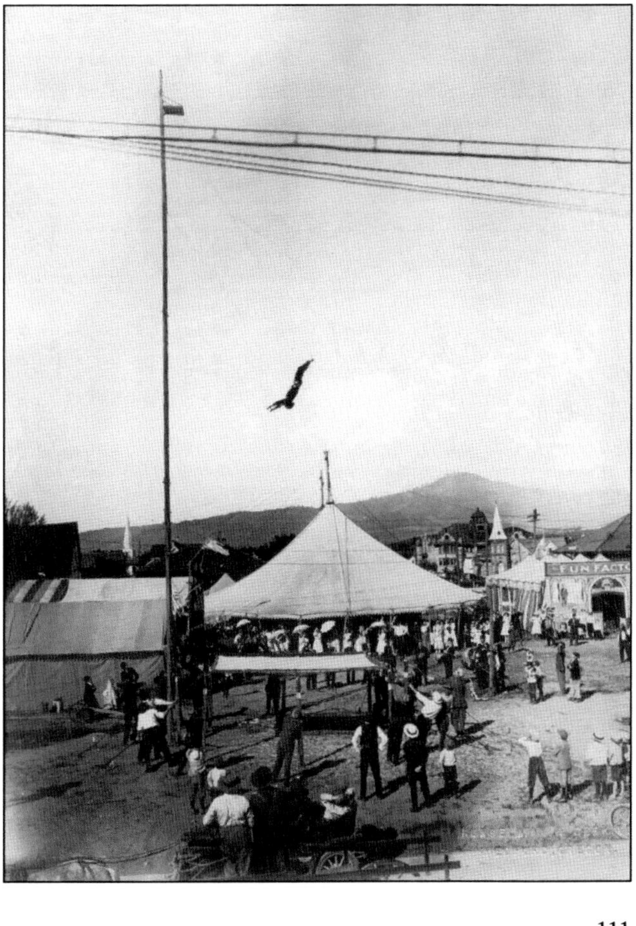

A critical moment at the circus in Medford, Oregon, was caught on camera. As early as the 1860s, circus companies brought wagons of wild animals and strange acts to the frontier. This "Fun Factory" circus show came through in 1912. People came from Talent and all over the valley to see it. (Jerry Kime.)

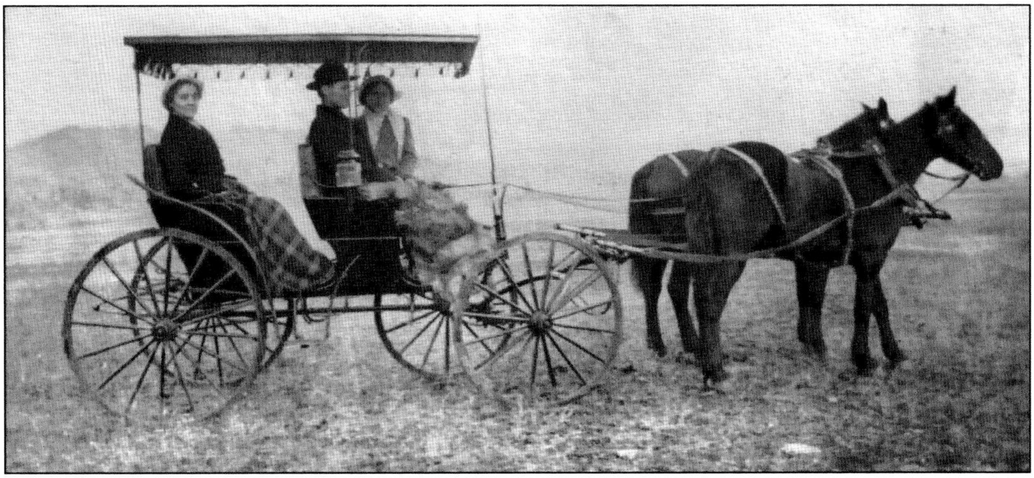

Albert Peachey was courting his girl, Margaret Daley, in a surrey with fringe on top. Margaret's mother, Alice Daley, chaperoned from the backseat. Couples must have developed a type of silent communication to show their feelings for each other when the chaperone was present.

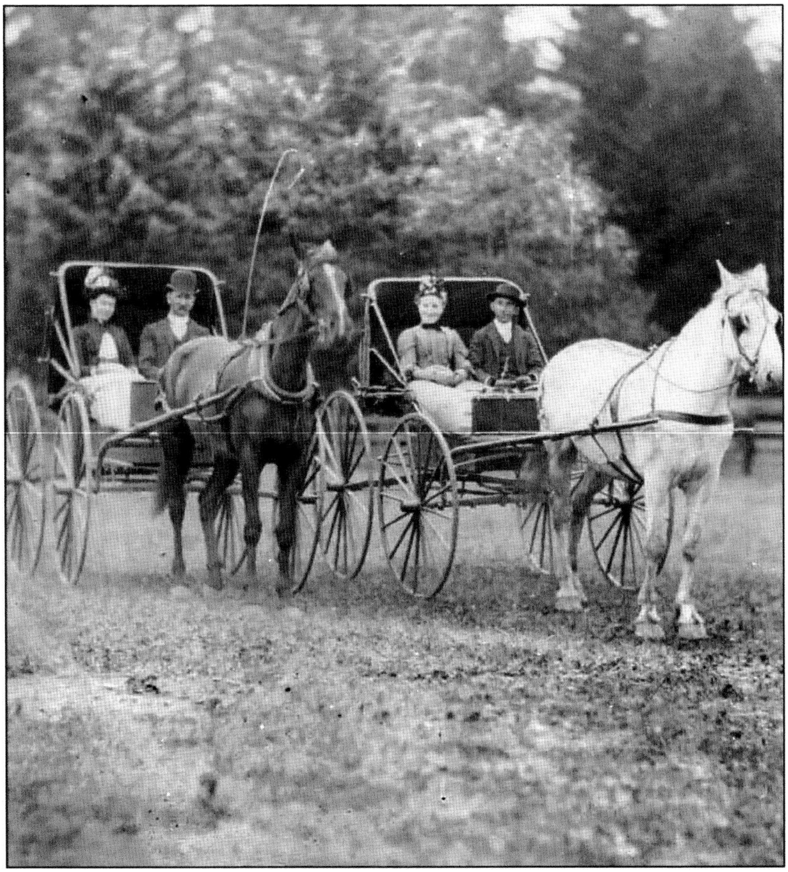

Etta, the daughter of Aaron and Martha Talent, married Lewis Lee in 1896. Lewis and his wife (on the left) were out for a buggy ride with their friends, the Denhams, in this old tintype. In 1906, Etta died at age 32 from a broken blood vessel, leaving her husband and little girl to mourn.

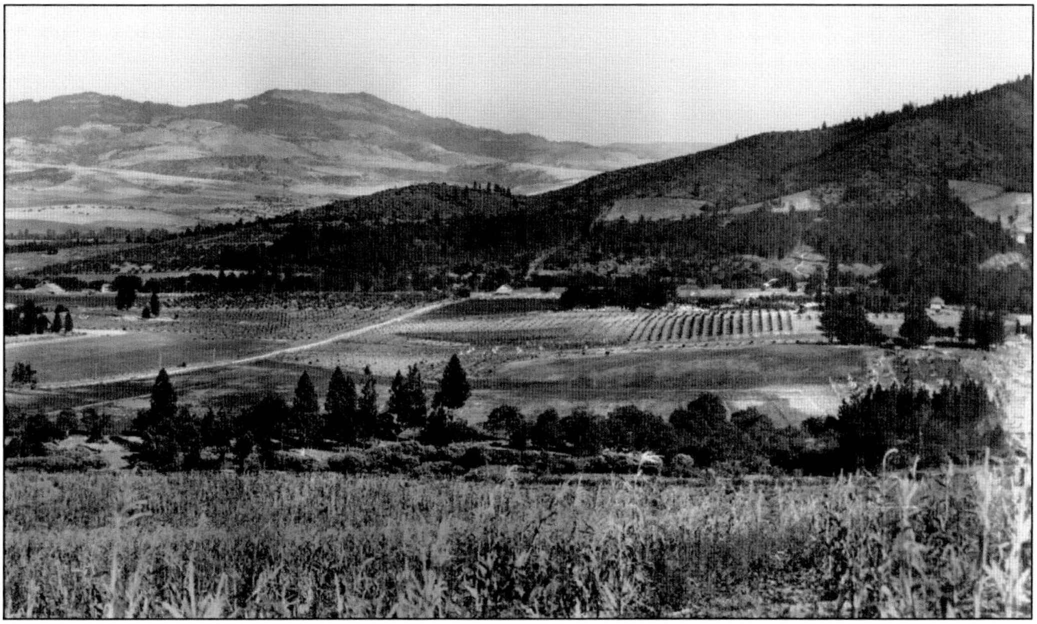

The Universal Mental Liberty Hall, pictured in the center of this photograph on the corner of Anderson Creek and Wagner Creek Roads, was a controversial place. From the time it was dedicated in 1885, local preachers called it "infidel hall" to draw attention to its freethinking and radical focus. Woman suffragists, socialists, populists, and atheists were allowed to debate and lecture there. (SOHS No. 20098.)

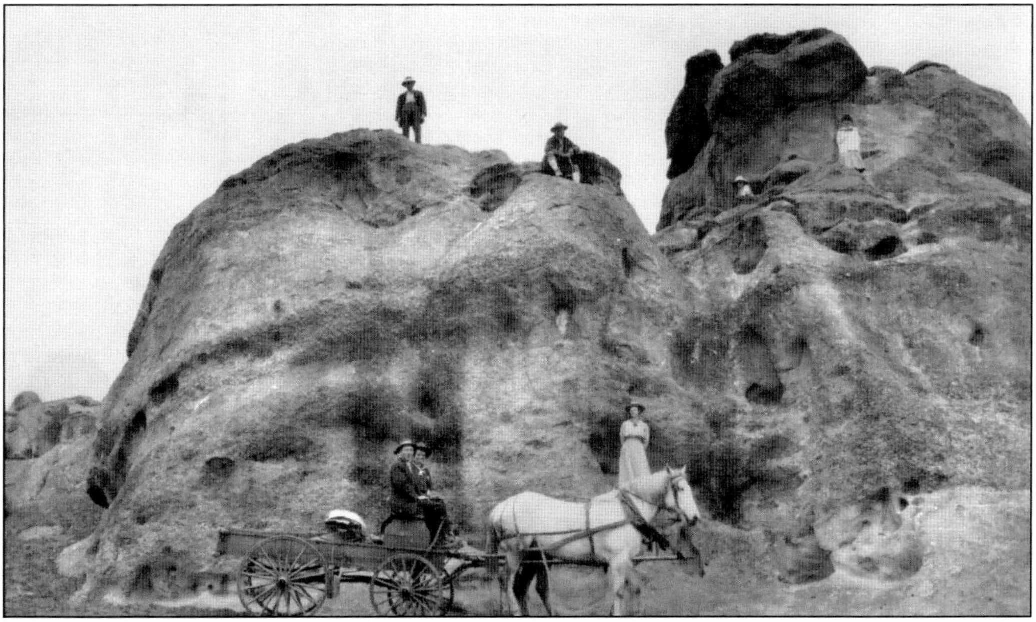

Mabel Kerby; Frank, Mary, and Fred Penniston; and Nina Hall paused on their way to Gaerky Falls near Grizzly Peak on the Cascade Ridge to let Charles Penniston take their photograph about 1912. A hiking club was formed about the same time called the "Grizzlies," whose members, like the Pennistons, kept photographs and scrapbooks to document their hikes all over southern Oregon. (Kerby family.)

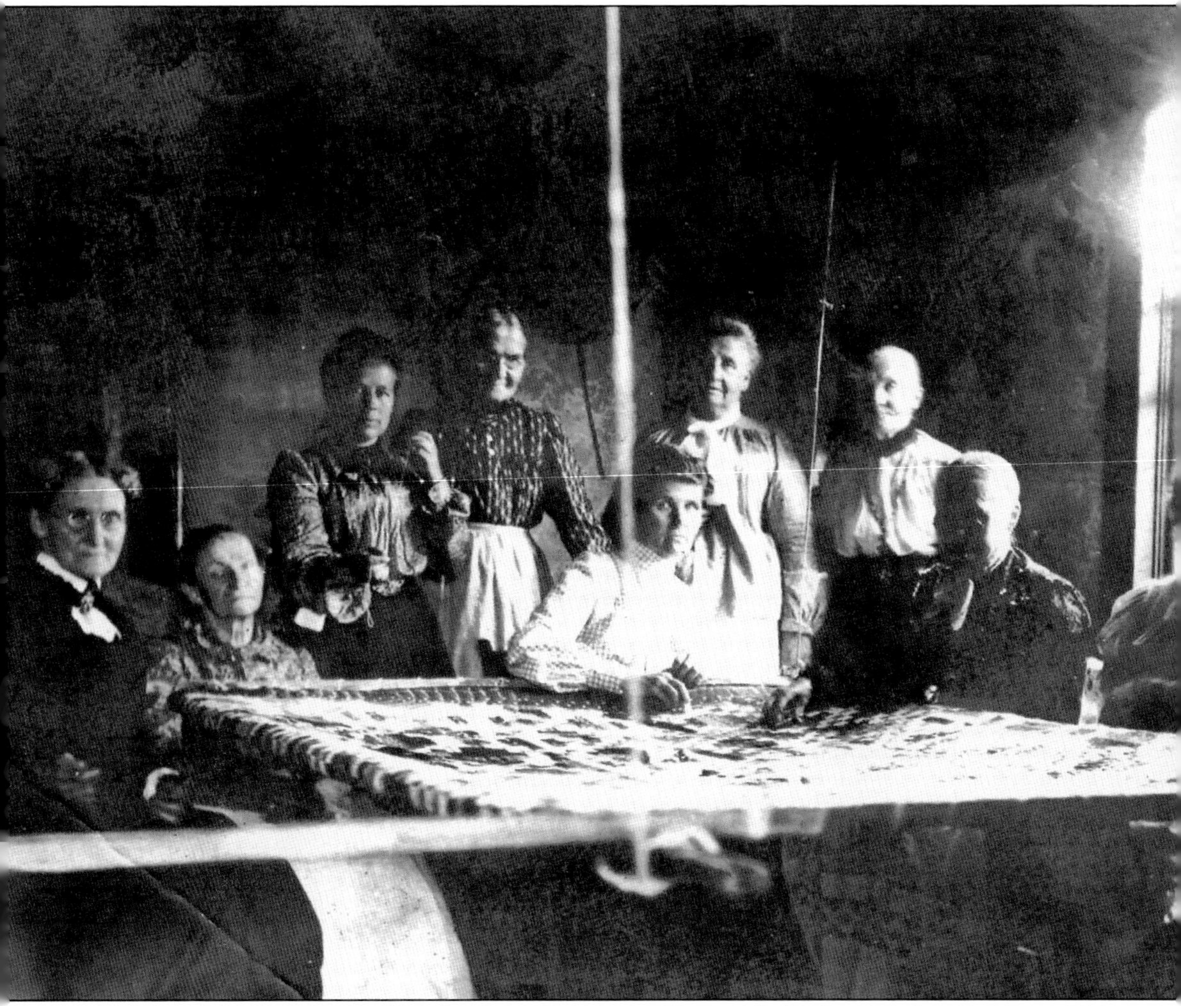

Quilts served artistic, social, and practical purposes. Women helped each other by saving fabric scraps, piecing them together, and then gathering together for the final quilting. The quilt in this image was hanging from the ceiling and could easily be retracted when the room was needed for other purposes.

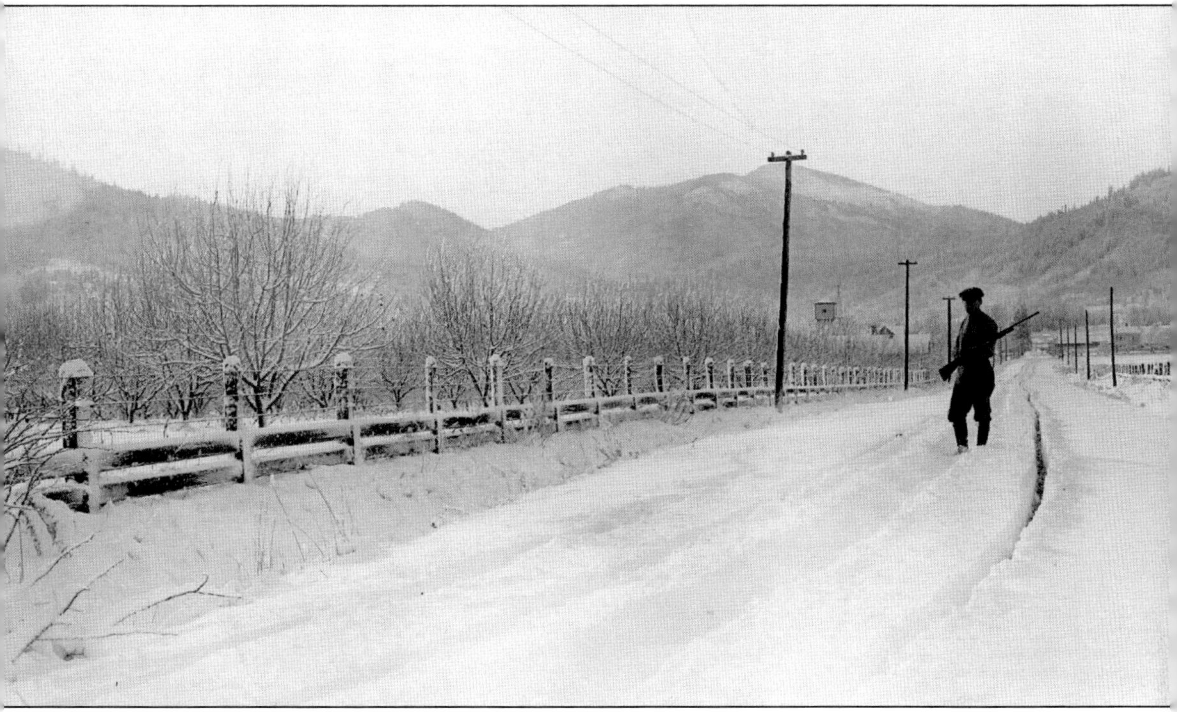

Standing in the middle of Wagner Creek Road, Charles Holdridge was photographed with his gun on his way to go hunting for easy targets that stood out against the snow. Old-timers say it used to snow more and stay longer than it does now. Charles would have agreed with them.

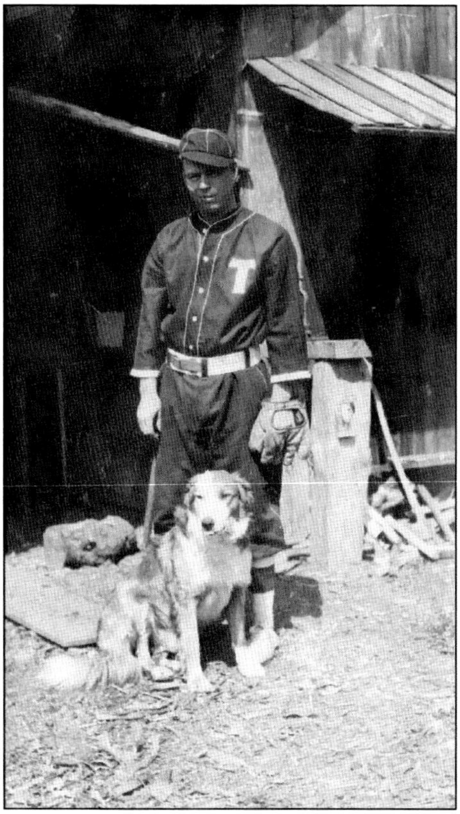

Charles Will Gleim Senior looks ready for anything in his oversized "T" for Talent baseball uniform. The faithful dog by his side also showed up in many of the Gleim photographs. (Harry Gleim.)

The Talent Baseball Team included Coach Garvin; Wayland Smith, pitcher; Will Gleim, center field; Moses Crawford, second base; Earl Beeson, right field; Harold Simpson, catcher; Chauncey Vance, first base; Doug Estes, third base; L. Thurber; C. Gardner; and Joe Barrett, batboy.

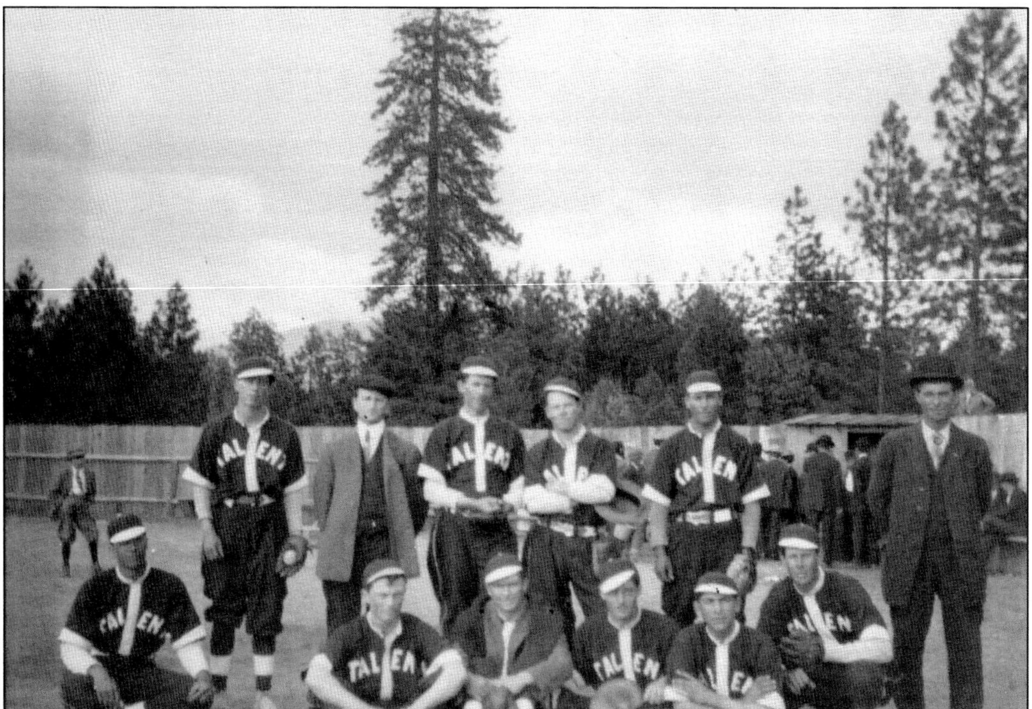

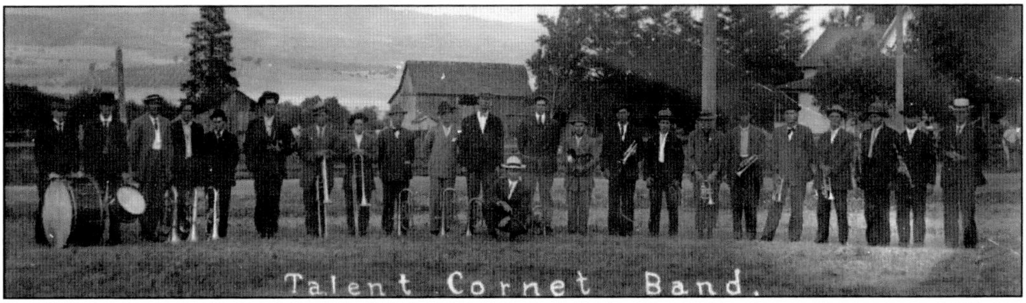

The Talent Cornet Band was funded by community leaders to enliven town events and fill the air with music. Band members included Everett Beeson, Clay Patterson, Frank Works, George Gardner, Orley Powers, Roy Estes, Delbert Goddard, and Jay Terrill. (Jerald McGrew.)

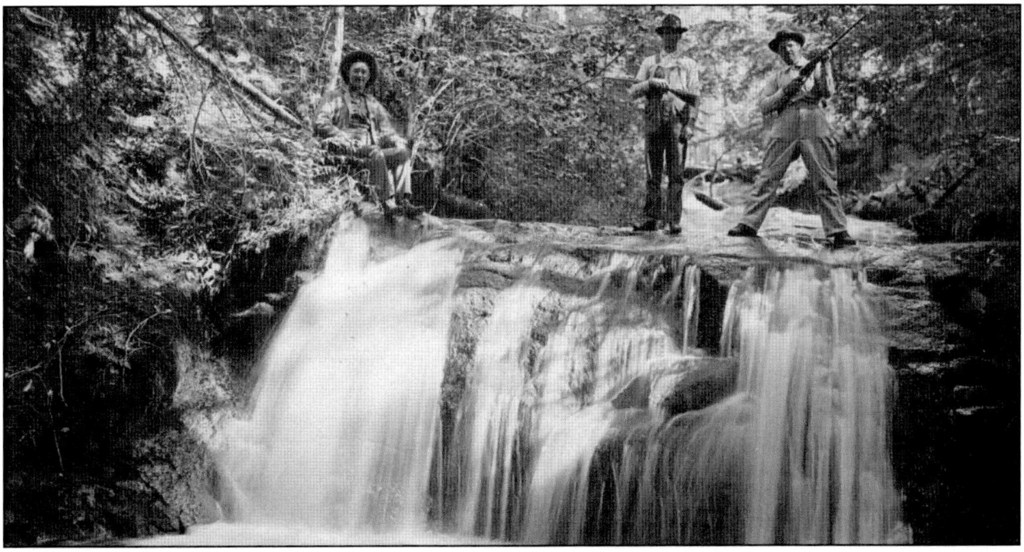

Guns are no longer allowed in Lithia Park in Ashland, but the Holdridge men from Talent took theirs with them on a hunting or hiking trip through Ashland Canyon about 1911.

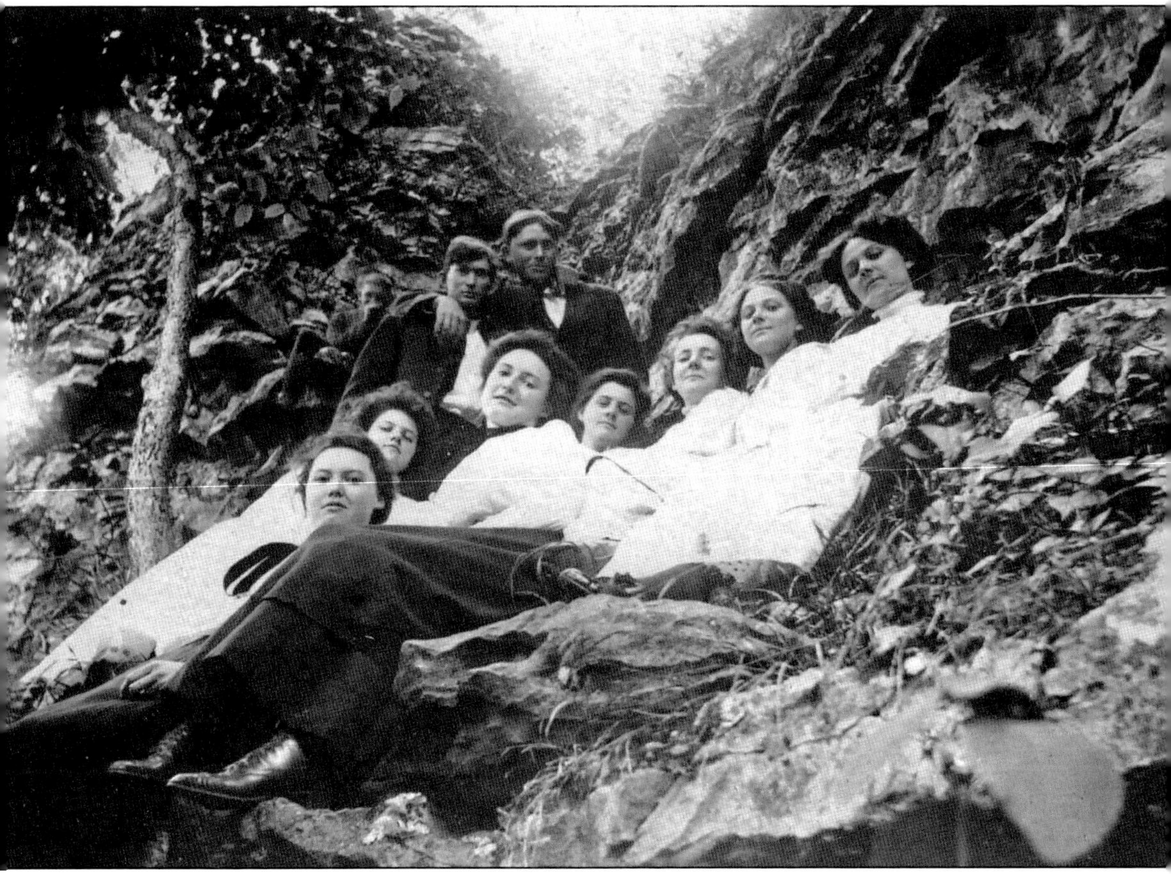

Three lucky young men got to go with seven young ladies on a hike. It is always a source of wonder how the women managed to look clean and the men so buttoned-down even when they are out in the dirt. (Jerry Kime.)

Women participated in sports but were limited in their movements by what would be considered today cumbersome clothing. This photograph comes from the Gleim family album. (Harry Gleim.)

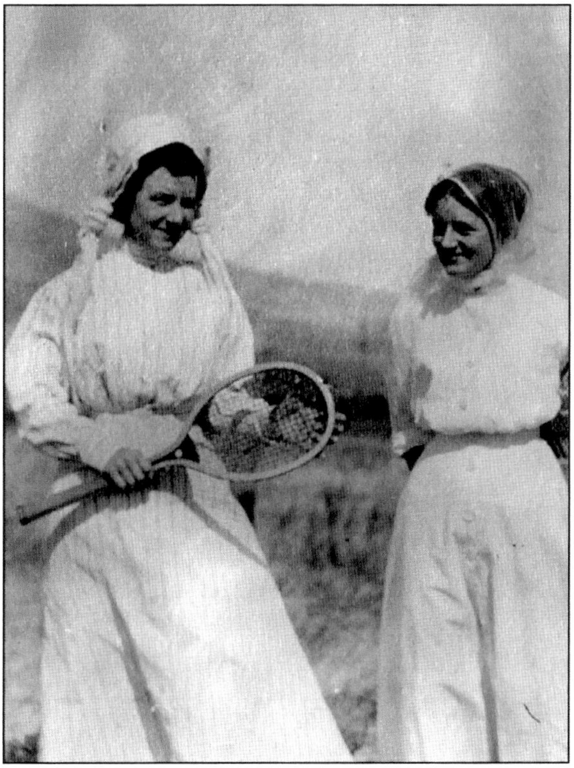

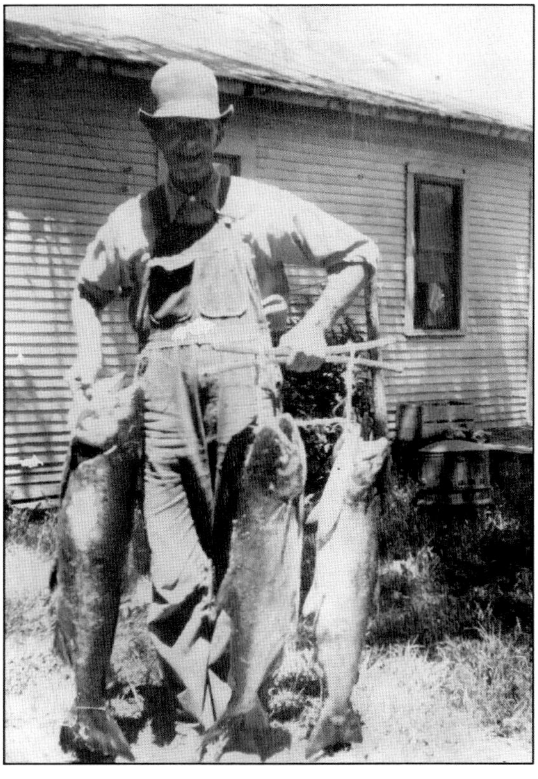

Frank Holdridge caught all the legally allowed fish he could in one day. His brother, Charles, the amateur photographer who took many of the pictures in this book, came over to record the exciting event.

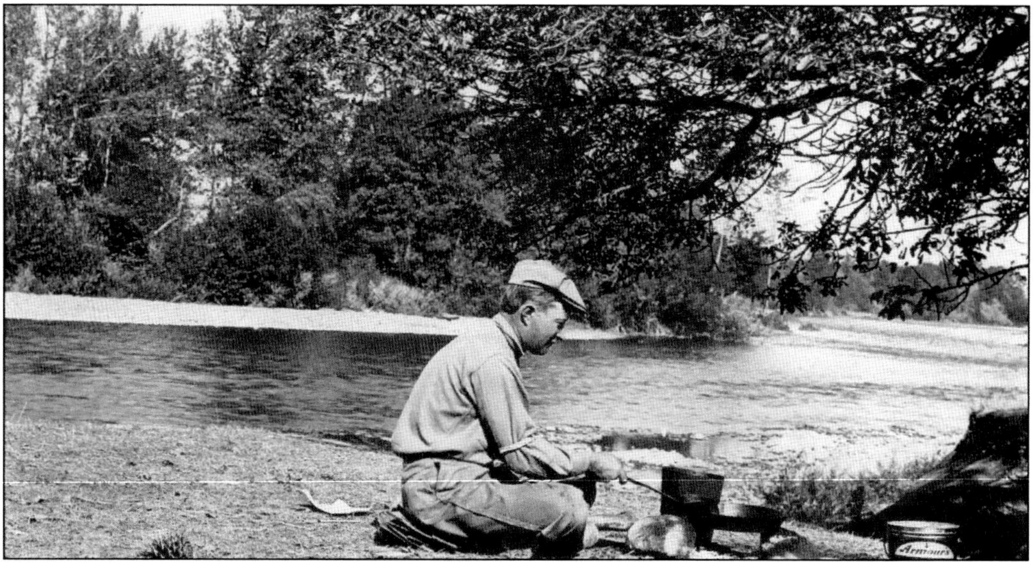

Bybee Bridge on the Rogue River was a popular camping spot. Charles Holdridge was seen with his camping and fishing gear near the bridge, where he could catch a salmon and fry it immediately for the best meal around.

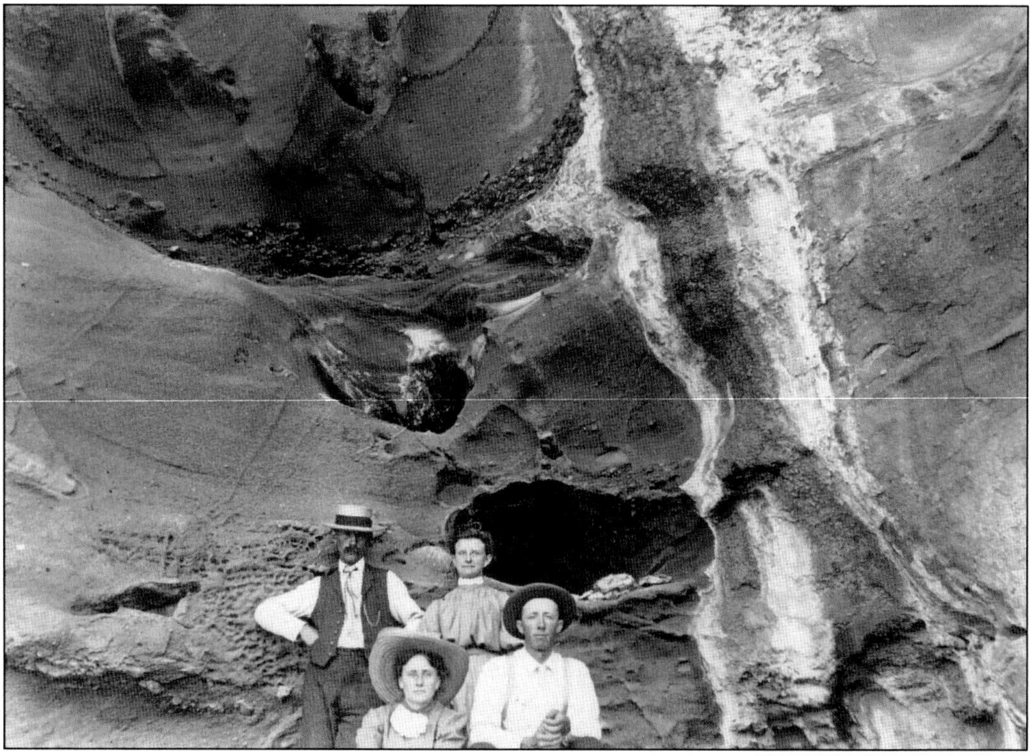

Delwin and Clara Hurley and Frank and Edith Holdridge took advantage of the summer weather and drove to the Oregon coast. It may have been their first view of the ocean, as they had all grown up in the Midwest before their families moved to Oregon.

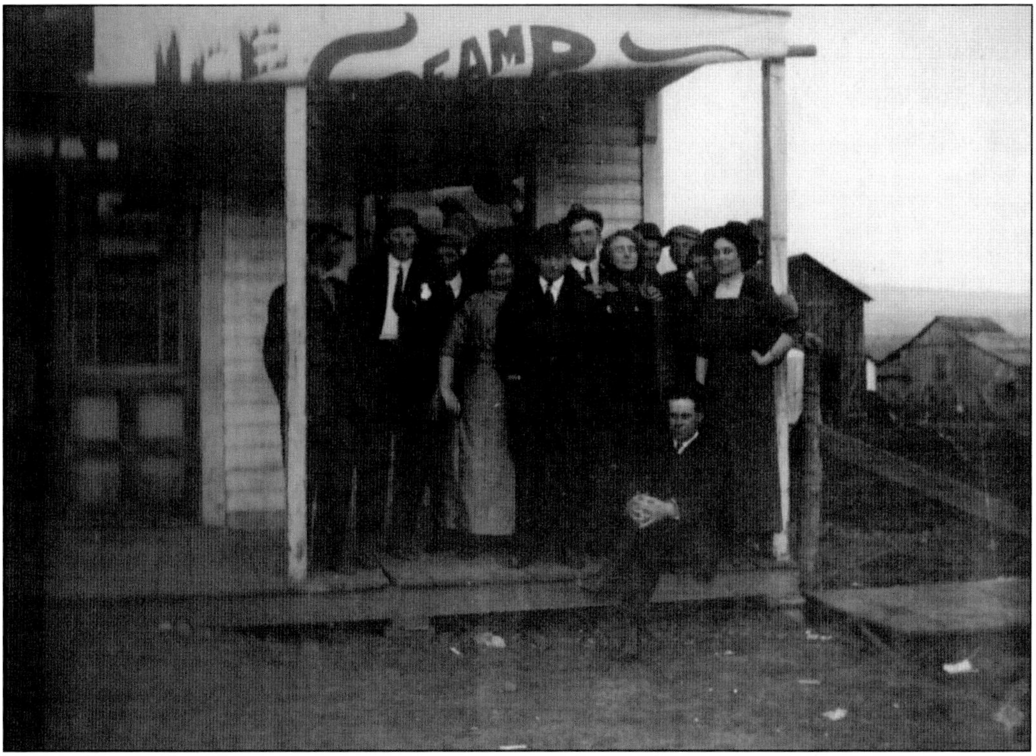

This image was found in a Gleim family album with several other photographs all taken on the same day. The people in the photograph walked all around Talent and took pictures, giving an informal view of the town. The Ice Cream Parlor was in the Hanscom Hall building on Talent Avenue. (Harry Gleim.)

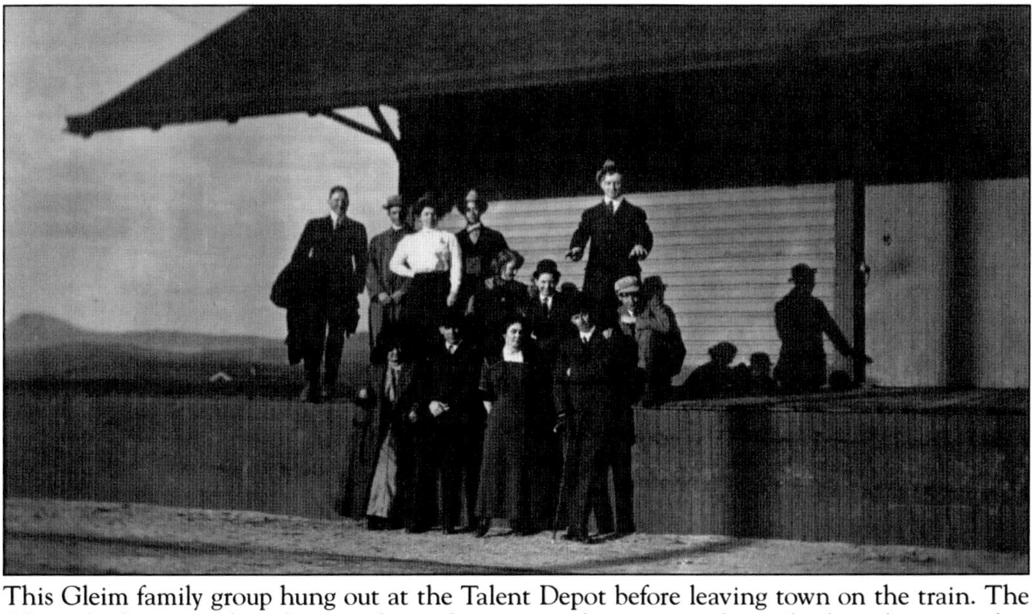

This Gleim family group hung out at the Talent Depot before leaving town on the train. The Gleims had ties to other places and were known to take occasional trips back to their cinnabar mine in Shafter, Texas, or to visit relatives in other states. (Harry Gleim.)

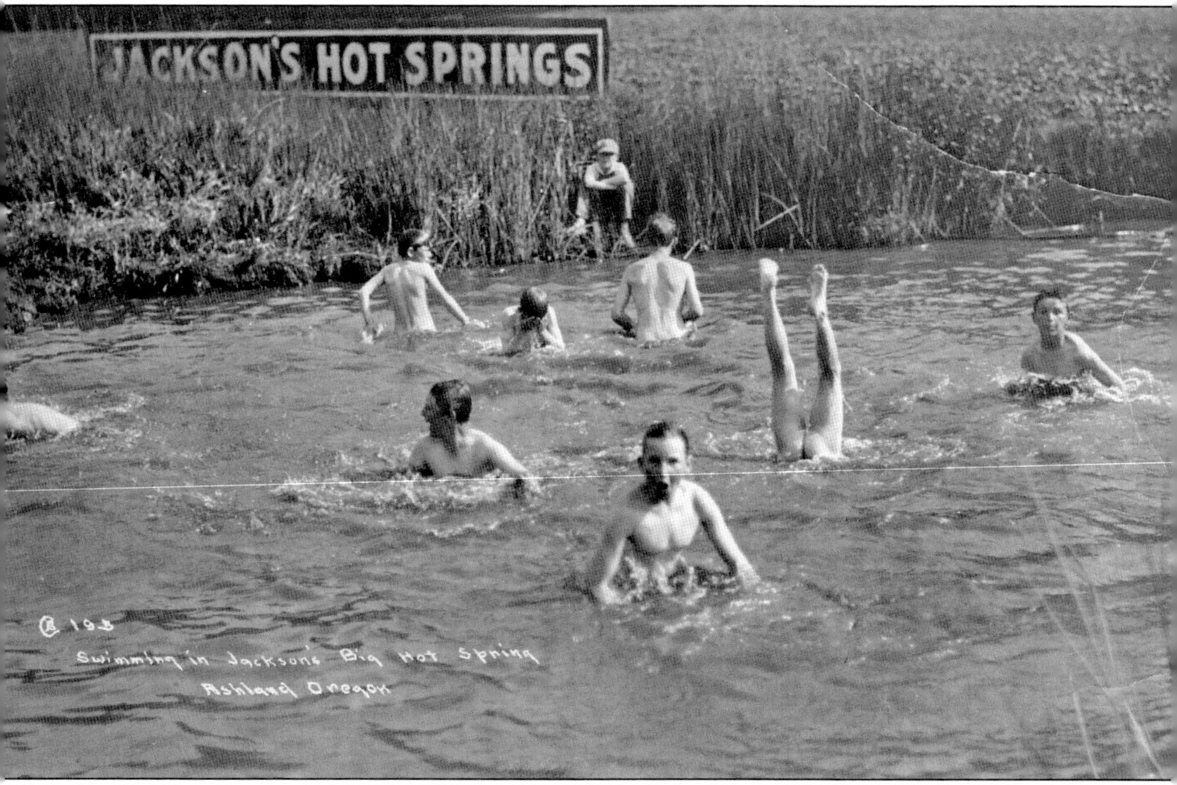

After chores were done, Talent boys could follow the railroad tracks toward Ashland down to Jackson Hot Springs and take a dip in the warm water. They were not just refreshing their bodies but getting completely caught up in the joy of play. (Kerby family.)

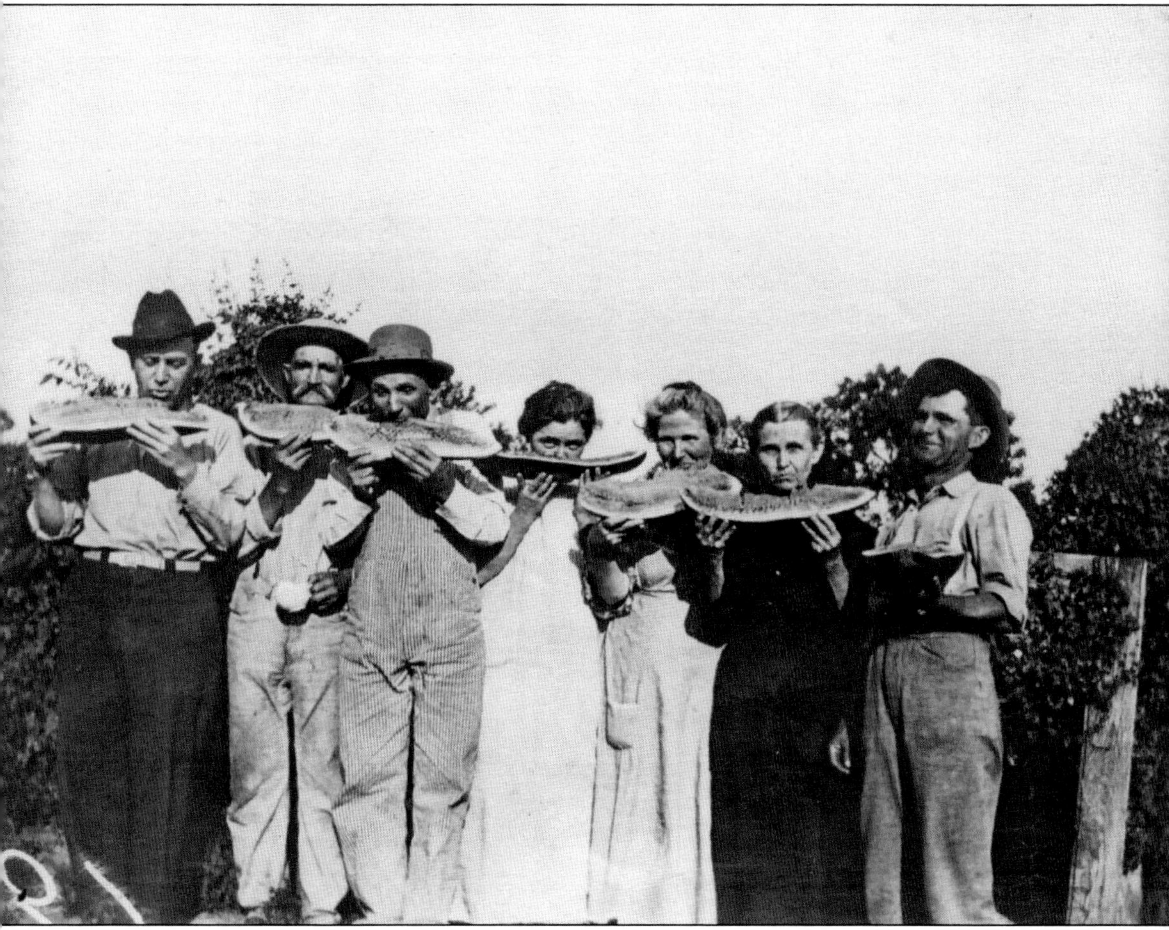

Talent was known for its watermelons and berries. The Quackenbush family lived just around the corner from the Fred Rapp watermelon patch and could enjoy all they wanted if they paid for it first. (Jerry Kime.)

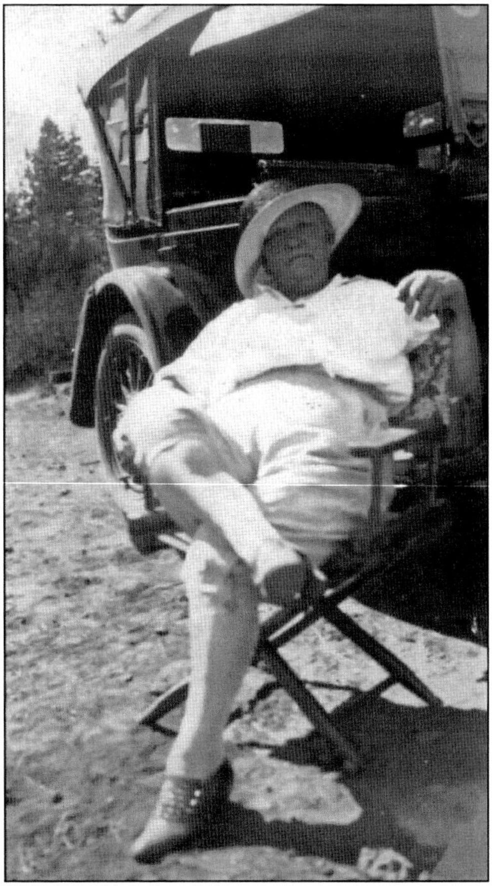

Marcia Coleman Robison looked quite commanding from her director's chair, with her great shoes and risqué knickers.

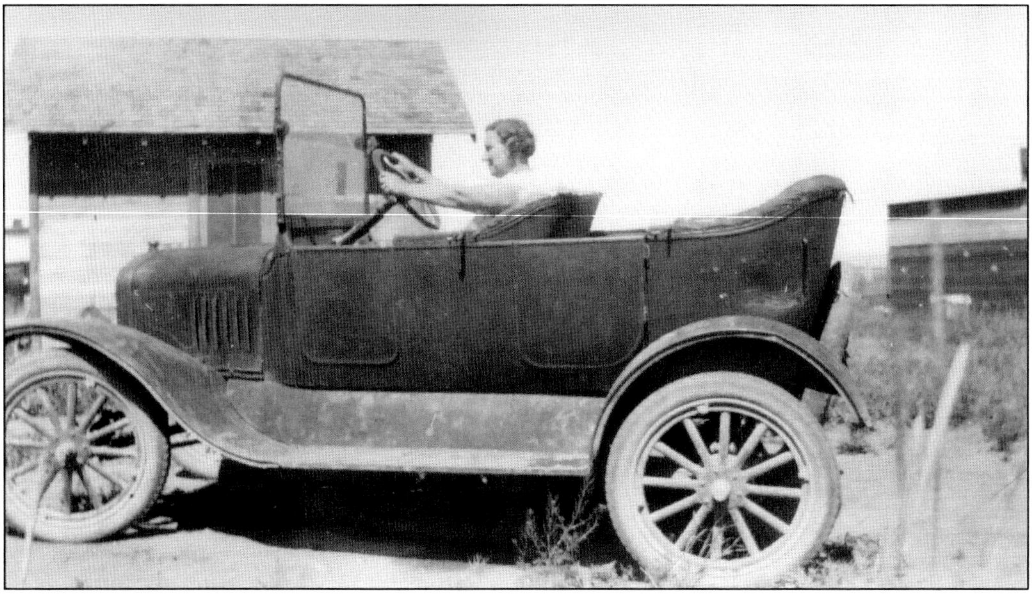

Ethel Childers in 1931 pretended to drive the old Model T. Judging from the weeds growing around the back wheels, it looks like it had been awhile since the car had moved.

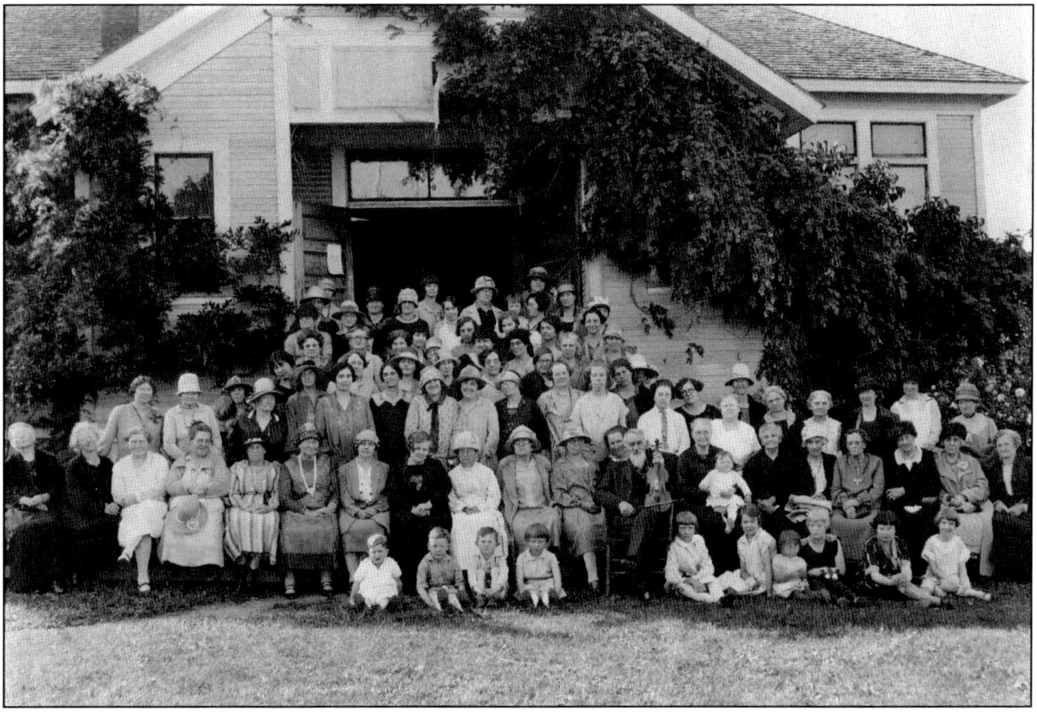

By 1927, when this photograph was taken, the Main Street schoolhouse was used as a Community Hall. A woman's group was meeting with one male guest, the fiddler, to serenade them.

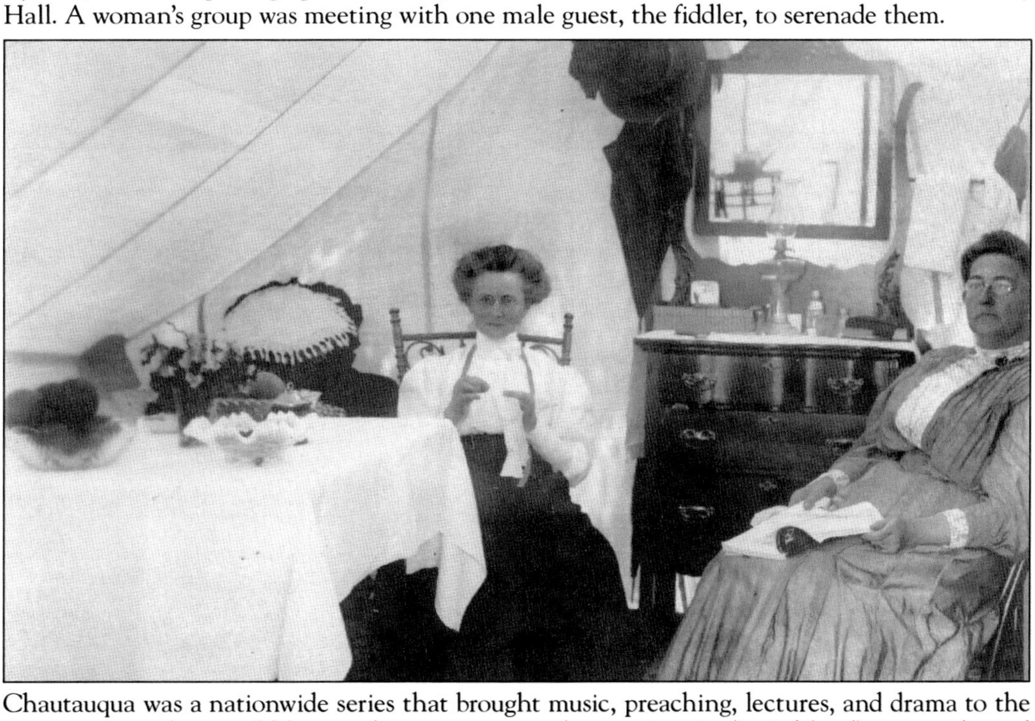

Chautauqua was a nationwide series that brought music, preaching, lectures, and drama to the country. Attendees would frequently camp out near the meeting site (in Ashland), as is evidenced in this 1908 photograph of Chautauqua goers. The ladies in this image had a tent set up complete with their furniture and their mending.

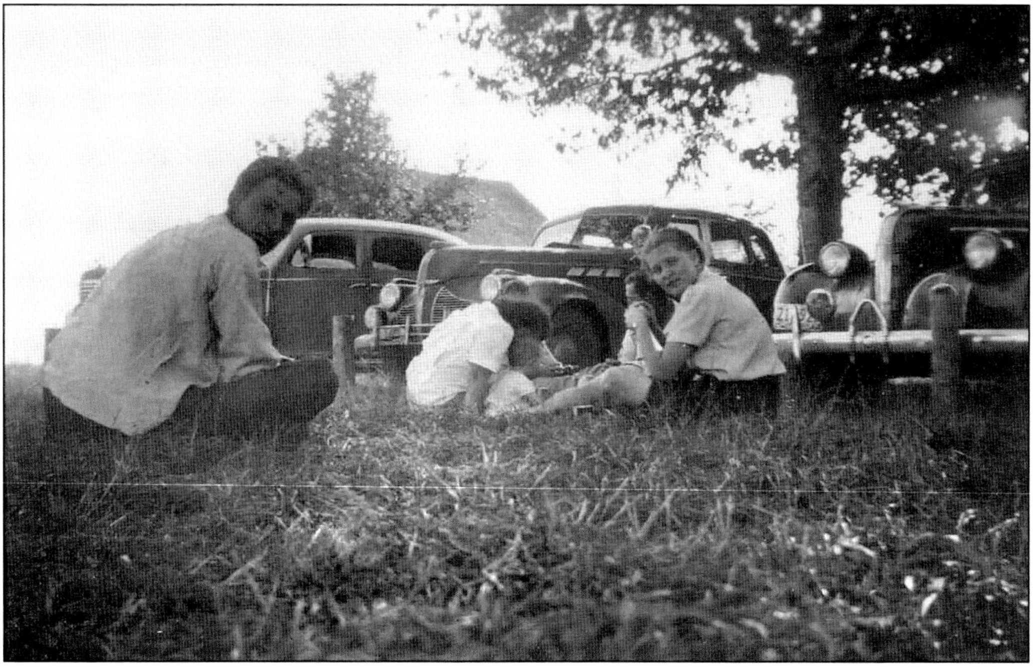

Photographs like this one lock in memories and highlight details that might otherwise be forgotten. Picnics on grassy hills never go out of style, as shown in this 1940s Kime family photograph. (Jerry Kime.)

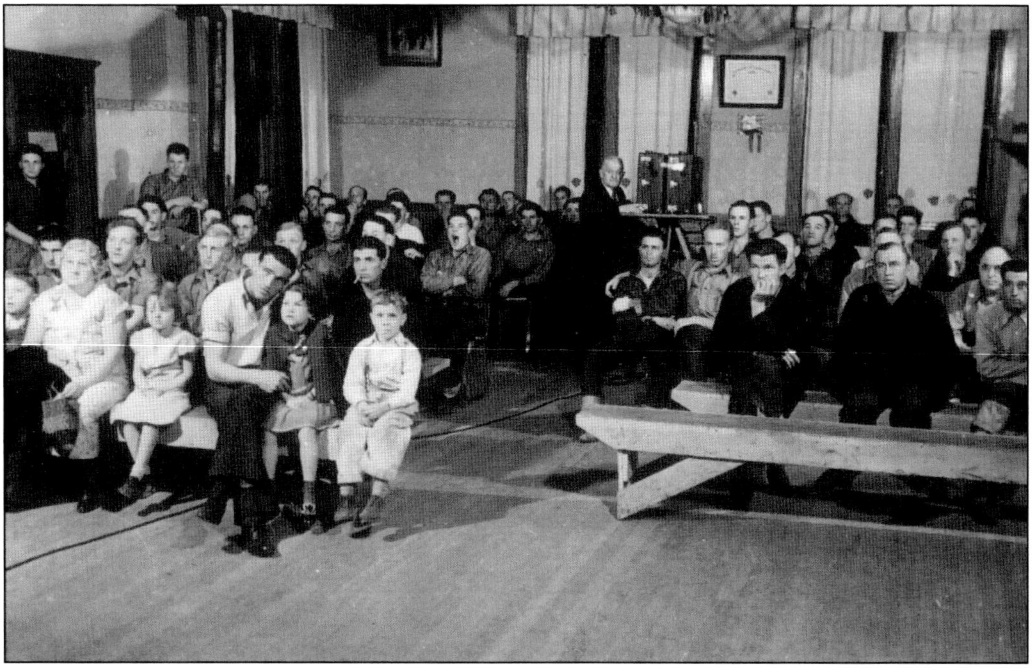

A special showing of the movie *Show Boat* for the Federal Relief Camp in the Talent Community Hall about 1936 brought in those looking for entertainment. Though Talent once boasted an opera house and now has the Camelot Theater, it has never had a formal indoor movie theater. (SOHS No. 11767.)

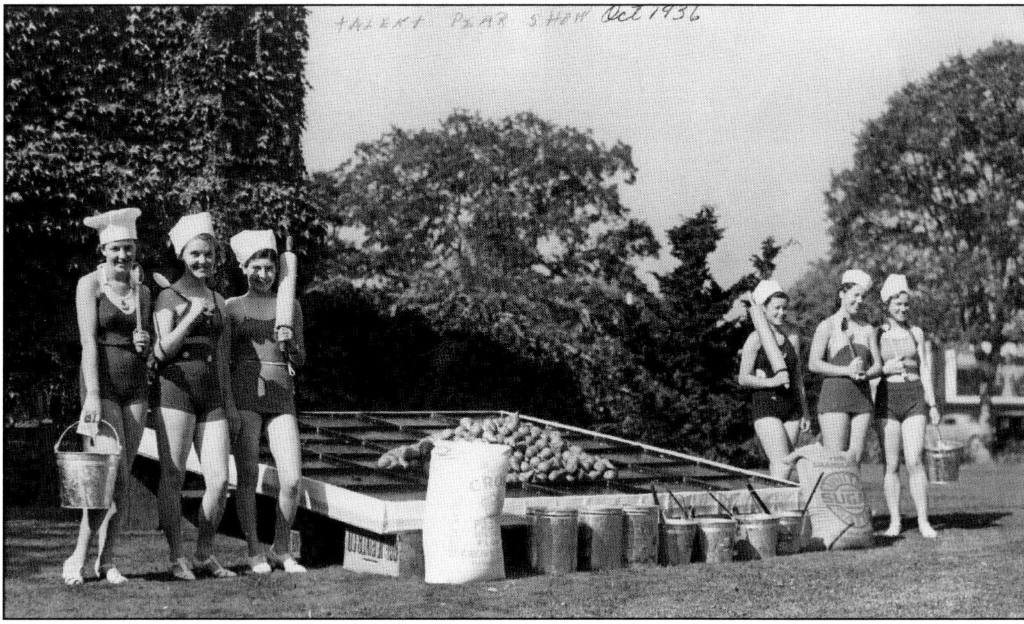

These young beauties posed as the bakers of the pear pie for the Pear and Tomato Show in Talent. They were, from left to right, Virginia Allen, Helen Combs, Margaret McDowell, "Queen" Opal Hill, Ella Mae Works, and Lola Conatser. It is doubtful that they actually made the pie and probable that this was the first and only time they got to parade around in their swimming suits on dry land.

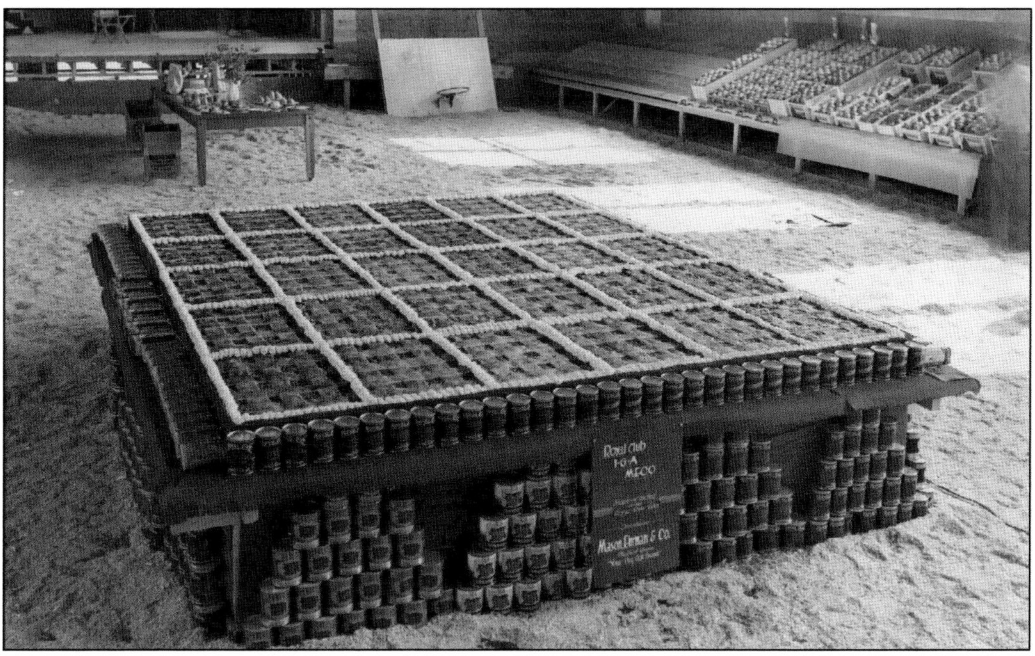

Everyone in town got a piece of this pear pie. It was the glory of the 1936 Pear and Tomato show in Talent, which preceded the famous Pear Blossom Days in Medford. The event also included a fat men's race, balloon shot put, a Kitten Ball Game between the Pear Pickers and the Tomato Pickers, and a pear-tossing contest.